MECHANICAL ENERGY
STORAGE TECHNOLOGIES

MECHANICAL ENERGY STORAGE TECHNOLOGIES

AHMAD ARABKOOHSAR
Associate Professor
Department of Energy Technology
Aalborg University
Esbjerg, Denmark

ELSEVIER

ACADEMIC PRESS
An imprint of Elsevier

Academic Press is an imprint of Elsevier
125 London Wall, London EC2Y 5AS, United Kingdom
525 B Street, Suite 1650, San Diego, CA 92101, United States
50 Hampshire Street, 5th Floor, Cambridge, MA 02139, United States
The Boulevard, Langford Lane, Kidlington, Oxford OX5 1GB, United Kingdom

Notices
Knowledge and best practice in this field are constantly changing. As new research and experience broaden our understanding, changes in research methods, professional practices, or medical treatment may become necessary.

Practitioners and researchers must always rely on their own experience and knowledge in evaluating and using any information, methods, compounds, or experiments described herein. In using such information or methods they should be mindful of their own safety and the safety of others, including parties for whom they have a professional responsibility.

To the fullest extent of the law, neither the Publisher nor the authors, contributors, or editors, assume any liability for any injury and/or damage to persons or property as a matter of products liability, negligence or otherwise, or from any use or operation of any methods, products, instructions, or ideas contained in the material herein.

Library of Congress Cataloging-in-Publication Data
A catalog record for this book is available from the Library of Congress

British Library Cataloguing-in-Publication Data
A catalogue record for this book is available from the British Library

ISBN 978-0-12-820023-0

For information on all Academic Press publications
visit our website at https://www.elsevier.com/books-and-journals

Publisher: Brian Romer
Acquisitions Editor: Lisa Reading
Editorial Project Manager: Sara Valentino
Production Project Manager: Prem Kumar Kaliamoorthi
Cover Designer: Christian Bilbow

Typeset by SPi Global, India

Working together
to grow libraries in
developing countries

www.elsevier.com • www.bookaid.org

Contents

6. Pumped heat storage system   125

7. New emerging energy storage systems   149

8. Conclusion   177

CHAPTER ONE

Classification of energy storage systems

Ahmad Arabkoohsar
Department of Energy Technology, Aalborg University, Esbjerg, Denmark

Abstract

In general, energy can be stored with different mechanisms. Based on the mechanism used, energy storage systems can be classified into the following categories: electrochemical, chemical, electrical, thermal, and mechanical. These methods are explained in the sections that follow.

1.1 Electrochemical

The two main categories of electrochemical energy storage systems are batteries and supercapacitors.

1.1.1 Batteries

Batteries are indeed the most straightforward and popular method of storing electricity in existing energy systems. Batteries, for which there are several subcategories, offer the most interesting efficiencies among all the other categories of energy storage. It can also be claimed that this category is more mature than most of the others as batteries are broadly commercially available, from the very small size of batteries for toys to the medium scale of renewable power plants. The most well-known battery technologies are lead-acid, nickel-cadmium (NiCad)/nickel-metal hydride (NiMH), lithium-ion, metal-air, sodium-sulfur, sodium-nickel chloride, and flow batteries. As this energy storage category is not the main focus of this book, these electrochemical technologies are very briefly discussed. The two general drawbacks of batteries, regardless of their subcategory, are their low energy density and their high cost (compared to other solutions) [1].

1.1.1.1 Lead-acid

This is the first battery technology, which emerged about 130 years ago. It is more broadly employed than any other batteries as well. Such batteries can be used for either stationary or mobile systems from very small scales to

Mechanical Energy Storage Technologies
https://doi.org/10.1016/B978-0-12-820023-0.00001-8

medium scales. Today, many electrical vehicles, standalone photovoltaic units, wind farms, and so on are taking advantage of this type of battery. Lead-acid batteries' useful lifetime can be in the range of 6–15 years with about 1500 cycles at 80% discharging depth. High efficiency of up to 90% could be expected from lead-acid batteries over a full charging-discharging process. This is, however, a matter of discharging rate as well, so that a milder rate of discharge would result in a better efficiency, while a fast discharging rate ruins the battery performance to efficiency levels as low as 50% [2].

1.1.1.2 NiCad/NiMH

Compared to lead-acid batteries, NiCad/NiMH batteries offer better energy density and a greater number of operating cycles. Among these two nickel-based battery types, NiCads are able to operate quite well at very low ambient temperatures. In larger sizes, NiCads have about the same performance level as that of lead-acid batteries. Due to their comprising the toxic material of cadmium, NiCads are only used for stationary applications today. Alternatively, having almost all the positive aforementioned features of NiCads, NiMH batteries offer a better energy density but far lower maximum nominal capacity than NiCad batteries. NiMH batteries are broadly used in hybrid vehicles today [3].

Lithium-Ion: This battery type is of great interest today and the rate of deployment of its market is probably the fastest among all other storage solutions in this category. Today, more than 50% of small portable devices use this type of battery. The energy density of lithium-ion batteries is much better than that of NiCad/NiMH batteries. They also come in high capacities, offer low self-discharge rates, and require little maintenance. Despite these advantages, lithium-ion batteries suffer from high cost, immature state-of-practice, and aging even when not in operation [4].

1.1.1.3 Metal-air

Metal-air batteries have very large energy density of around 12 kWh/kg, mainly due to the open configuration of the storage that uses air as the reactant, which makes them useful in the automotive industry. However, metal-air batteries are not commercially available yet. These batteries are among the very likely solution candidates for future electrochemical energy storage for a wide range of uses, from electric vehicles to power plants. To get to that point, however, challenges related to these batteries' metal anodes, air cathodes, and electrolytes need to be addressed [5].

Sodium-Sulfur: These batteries, which are of molten-salt battery type, use molten sulfur and molten sodium as cathodes and anodes, respectively. Since sodium-sulfur batteries operate at high temperatures (greater than 300°C), they require an auxiliary heater in many cases. The roundtrip efficiency of such batteries can reach up to 90% with a minimum energy density of about 150 Wh/kg. Their expected lifetime is about 10 years. These batteries can be found in the range of 1–100 MW. Thus, they are typically used as the main energy storage units of renewable power plants [6].

1.1.1.4 Sodium-nickel chloride
Similar to the sodium-sulfur battery, the sodium-nickel chloride battery has sodium as the anode, while it has an electrode consisting of both nickel and sodium chloride as the cathode. This battery is also a high-temperature kind (270–350°C) and needs an auxiliary heater. The roundtrip efficiency, energy density, and maximum lifetime of this battery are 85%–90%, 100–120 Wh/kg, and 15 years, respectively. Battery size is in the range of 4–25 kWh, which makes it be suitable for a wide range of applications requiring from a few kilowatts to several megawatts, including residential uses, grid stability/peak shaving applications, energy storage for renewable power plants, and so on [7].

1.1.1.5 Flow batteries
A flow battery, in contrast to the aforementioned conventional battery, is a type of rechargeable electrochemical storage in which rechargeability is done by two chemical components dissolved in liquids. In conventional batteries, energy is stored as the electrode material, whereas in flow batteries, energy is stored as an electrolyte. There are three main categories of flow batteries: rodex, hybrid, and membrane-less flow batteries. Flexible layout, long lifetime (more than 20 years), instant recharging by replacing the electrolyte liquid, quick response, and no harmful emissions are the main advantages of such batteries. Their disadvantages are low energy density and low charge/discharge rates. Flow batteries an be found in capacities ranging from 25 kW to more than 100 MW [8].

1.1.2 Supercapacitors
Compared to rechargeable batteries, supercapacitors offer much lower energy densities (Wh/kg) and much larger specific power (W/kg). Therefore they have much shorter and simpler charging-discharging cycles (without being subject to overcharging) compared to batteries. Supercapacitors offer a considerably longer lifetime (virtually unlimited) with hundreds of times larger charging-discharging cycles as well as broader charging-discharging temperature ranges. However, supercapacitors cost more per kWh and have greater self-discharging rates than batteries [9].

1.2 Chemical

While batteries are considered to be in the category of chemical energy storage due to the chemical basis of how batteries operate, this book defines chemical energy storage systems as a class of technologies that convert electricity to a form of potential energy carrier via chemical reactions. In other words, chemical energy storage systems are defined as those systems that employ any source of surplus electricity from a renewable power plant to drive a chemical reactor that might produce any type of fuel. These methods are also known as Power-to-X methods. The possible technologies in this category are power-to-gas, power-to-liquid, and power-to-biofuels. The produced environmentally friendly fuels can then be simply transported through existing oil/gas pipelines or be used for driving conventional power plants or vehicles [10].

In these energy storage methods, via an electrolysis process, electricity separates hydrogen and oxygen from water to produce a gaseous or liquid fuel. Chemical energy storage systems are the most straightforward with the best energy conversion efficiency. The produced hydrogen can be simply used as a clean fuel itself. Alternatively, the produced hydrogen can be mixed with carbon dioxide, resulting in methane as another carbon–neutral synthetic gas. This process, which is called methanation, is indeed a power-to-gas process, but with more conversion losses than the first method. If the methane is converted to liquefied petroleum gas via synthesizing at very high pressures and low temperatures, it is a power-to-liquid process. If the hydrogen is mixed with biogas generated in a biogas production plant in order to upgrade the quality of the biogas (e.g., from a wood gas generator), this process is called power-to-biogas [11].

1.3 Electrical

There are two main methods for electrical energy storage approaches: capacitors and superconducting magnetics.

1.3.1 Capacitors

While both batteries and capacitors are used for electrical energy storage, they are fundamentally different. A capacitor stores energy in an electric field, while a battery stores electricity in chemical form. In fact, a capacitor is a passive two-terminal electrical system with at least two plates as electrical conductors that are separated by an insulator as the dielectric, storing electricity in the electrostatic field between the two plates. Conventional capacitors have a dielectric between plates instead of an electrolyte and a thin insulator as in supercapacitors.

There is a wide range of capacitor types that are broadly used, from very small scales in the electrical circuits of many electrical/electronic devices to large scales for power factor correction applications. Conventional capacitors offer very low specific power of less than 360 J/kg, while this parameter is far greater for conventional batteries [12].

1.3.2 Superconducting magnetics

Such a system stores energy in a magnetic field created by the flow of direct current in a superconducting coil that has been cooled to a temperature lower than its superconducting critical temperature. The most outstanding feature of a superconducting magnet is its ability to support a very high current density with vanishingly small resistance. Since the current densities are high, superconducting magnet systems are quite compact and occupy only a small amount of space. Superconducting magnetic systems offer roundtrip efficiency greater than 95%. Due to the energy requirements of refrigeration and the cost of superconducting wire, a superconducting magnetic system is used for short duration storage, such as for improving power quality. It also has applications in grid balancing [13].

1.4 Thermal

Thermal energy storage (TES) technologies, as the name implies, are mainly used for storing thermal energy in the form of heat for later use. However, storing surplus electricity in the form of heat has received special attention recently as well. This will be thoroughly discussed in Chapter 2 "Thermal Energy Storage Technologies."

Heat can be stored in different ways. The most straightforward method is sensible heat storage, but latent heat storage is also an option for some specific applications. Regardless of the category, thermal energy storage units can be used for short-term applications (within a few hours), mid-term applications (a couple of weeks or so), or long-term or seasonal storage (a few months or more). Thermal storages are usually very efficient. Depending on the difference between the storage medium temperature and the ambient temperature, the duration of storage, and the effectiveness of the insulation, thermal storage efficiency can reach up to 100% [14].

1.4.1 Sensible thermal energy storage

This category of TES includes those heat storage units in which the stored heat flow leads to an increase in the temperature of the storage medium. For example, a hot water storage tank that gets hotter as its being charged and cooler when it is discharged. Water is the most popular storage medium for low-temperature applications in this category. Packed beds of stones or alternative materials are used for high-temperature thermal storage applications [15].

1.4.2 Latent thermal energy storage

Unlike in sensible heat storage, in latent thermal energy storage units the medium of heat storage goes through a phase change process rather than a temperature change. As the latent heat of heat storage materials is usually considerably high, the energy storage density of such storage units is quite high making them appropriate for large-scale applications. Molten-salt heat storage of solar thermal power plants is an example of such heat storage systems in which the gathered solar thermal energy is used to gradually melt the solid-state salt in the storage unit with a high melting temperature suitable for the power plant. Then, the molten salt is used to vaporize the pressurized water of the power block of the plant in its discharging phase. In this phase, the molten salt releases energy (its latent heat) and gradually goes back to its initial solid state. There is a wide range of phase change materials broadly used for various heat storage applications from very low temperatures to very high temperatures [16].

1.5 Mechanical

Mechanical energy storage systems are those energy storage technologies that convert electrical energy to a form of storable energy flow (other than electricity) when charging to reclaim it for electricity production (or co- and tri-generation) over a discharging phase. In most of these technologies, the surplus electricity of a renewable power plant is used to drive a mechanical system. When the power plant demands compensation, the energy storage unit comes into operation to generate the required amount of electricity. These systems are emerging due to the lack of appropriate and cost-efficient large-scale energy storage solutions in the other categories of energy storage. The literature gives information about several different mechanical energy storage systems, each of which has its own advantages and limitations.

1.5.1 Pumped hydroelectric storage

This is probably the oldest mechanical energy storage system that is still among the most efficient ones. The idea behind this storage technology is that the available surplus electricity to be stored is used for pumping water downstream from a large water dam upstream, so that it gets the potential energy for producing power flowing through a water turbine when required. This system is quite simple in terms of operation, does not require sophisticated technologies of system components, and may result in roundtrip efficiency of about 85%. The technology is now well developed and available in the market in many locations around the world. However, the problem with this technology, besides its high cost of investment, is the very special geographical conditions required for building large water dams. In Europe, countries like Norway with lots of hills and mountains are much more appropriate for such plants than countries with flatlands like Denmark or the Netherlands [17].

1.5.2 Flywheel

Flywheel energy storage is a smart method for storing electricity in the form of kinetic energy. The idea behind this technology is that the surplus electricity to be stored drives a motor that spins a flywheel thousands of rounds per minute to store kinetic energy. The flywheel moves easily because of being levitated in an evacuated chamber with magnets and highly efficient bearings. The stored kinetic energy is the momentum of the flywheel and can actuate an electricity generator as another part of the system to produce power. Low maintenance costs, a long expected lifetime, fast response, and roundtrip efficiency of about 90% are of the main advantages of flywheel systems. The main disadvantages are high cost, self-discharge risk, and appropriateness for smaller capacities only (from 3 kWh to 130 kWh) [18].

1.5.3 Compressed air energy storage

A compressed air energy storage (CAES) system is another promising mechanical electricity storage technology. The idea of this storage system is to utilize excess electricity to generate compressed air at very high pressures via driving compressors and then store the generated compressed air in a vessel or chamber to be used later for electricity production. For power generation, in the discharging step, the compressed air is reclaimed from the storage vessel to be expanded through air turbines. The turbines are naturally coupled to an electricity generator. This technology has a wide range of possible configurations, each of which has its own advantages and disadvantages. The system can be sized for even super large plants (e.g., capacity for hundreds of megawatts) [19]. The best possible roundtrip efficiency achievable by CAES technologies is about 80%. In CAES systems, multi-stage compressors and expanders are employed and the heat generated in the airflow while being compressed is stored to be used in the discharging mode for preheating the compressed airflow before expansion [20]. Then, the airflow is heated up to way higher temperature by auxiliary heaters. The main problem of this technology is the compressed air storage vessel. In large-scale uses, the vessel is supposed to be an underground cavern, which is costly and not always possible due to geographic limitations [21].

1.5.4 Pumped heat electricity storage

Pumped heat electricity storage (PHES) is a new generation of mechanical energy storage systems that has not been commercialized yet. However, thorough initial research and development activities have been accomplished and very promising outcomes have been achieved. In PHES the electricity flow to be stored is used to drive a compressor that compresses airflow and, in this way, greatly increases the temperature of the airflow. Then, the generated heat is stored in a high-temperature thermal energy storage bed. The compressed airflow, after giving its temperature to the heat storage medium and being cooled, is expanded via a turbine through which it is totally depressurized and cooled to very low temperatures. The subcooled depressurized airflow then goes to another thermal energy storage bed, which is a cold storage. Passing through the cold storage, the airflow gives its cold energy to the storage medium and gets back to its initial temperature. Then, the cycle is repeated until the hot storage is fully charged and the cold storage is fully discharged (has reached the uniform minimum temperature of the cycle). In the discharging mode, the cycle gets reversed so that the compressed airflow initiates from the hot storage. Thus, flows at high temperature through the turbine are expanded and generate work. Next, the expanded medium-temperature airflow passes through the cold storage to be cooled further down and finally enters the compressor to be compressed and heated again [22]. This system offers roundtrip efficiency of 85% in the most advanced and optimally designed configurations. However, as mentioned, the main challenge of PHES is that it is not a well-developed system, particularly with respect to the special compressors and turbines required [23].

1.5.5 New emerging storages

In addition to the aforementioned mechanical energy storage systems already introduced to the market and literature, there are a number of innovative electricity storage technologies that have potential to be popular in the future.

1.5.5.1 High-temperature heat and power storage

Based on the same idea as that of PHES (i.e., storing electricity in the form of high-temperature heat), the concept of high-temperature heat and power storage (HTHPS) was designed and presented to the literature in 2016 [24]. This storage technology uses simple electrical coils for converting available surplus electricity to high-temperature heat (around 700°C) and stores it in a packed bed of stones [25]. Then, for producing electricity from the stored energy in the discharging phase, the high-temperature heat is used to drive a conventional power block (either a steam cycle or a gas turbine cycle). The discharged heat from the heat supply unit of the power block (the boiler of the steam cycle or the heat supply chamber of the gas turbine cycle) is at high temperature and can be used for medium-temperature heating applications such as district heating supply. Thus, the HTHPS system is a co-generation technology appropriate for any large-scale capacity. The power-to-power efficiency of this technology may be in the range of 28%–35% and its power-to-heat efficiency can be in the range of 45%–65%, depending on the technology and design of the power block [26].

1.5.5.2 Subcooled CAES

Another category of multi-generation electricity storage systems is the newly emerged concept of subcooled CAES (SCAES), which has shown very interesting performance for large-scale applications and integration of different energy sectors [27]. As the name implies, this technology is a modified configuration of CAES technology in which the heat gathered in the compression phase is not stored but rather is directly supplied for district heating support. The compressed airflow before the expansion is not heated at all so that it cools to a very low temperature in each expansion phase. Then, it has the capacity to support cold for district cooling or industrial cooling uses. Thus, the system is a trigeneration technology that is appropriate for large capacities [28]. A SCAES unit may offer electricity-to-electricity,

electricity-to-heat, and electricity-to-cold efficiencies of 30%, 90%, and 30%, respectively. The overall coefficient of performance of a SCAES system is about 1.5 [29].

1.5.5.3 Gravitational storage

This system works using gravitational potential for storing electricity. In the charging mode, the excess electricity is used to dislocate a massive mass towards height (opposite gravitational direction) to create a huge potential for work production. Then, when needed (i.e., discharging), the mass due to the gravity effect moves downward and creates the conditions of producing work (electricity). A general configuration of this technology includes a very large hole in which a super heavy mass has been located. When there is surplus electricity, pumps are used to pressurize water below the mass and move the mass upward. This process continues until the mass is located at the top of the hole, that is, at the highest potential of the system. Then, in the discharging stage, the mass due to the gravity effect moves downward pushing the pressurized water below that through a water turbine that drives an electricity generator. This system, which might be considered as an improved generation of hydropower storage technology, is simple and does not require dams or special geographical needs. Thus, like pumped hydropower, a very important feature of this technology is its very fast response time to the power need or surplus available power to be stored. It takes only milliseconds to start the operation and just a few seconds to reach the nominal load. The roundtrip energy efficiency of this technology is expected to be about 90% [30].

References

[1] Y. Yang, S. Bremner, C. Menictas, M. Kay, Battery energy storage system size determination in renewable energy systems: a review. Renew. Sust. Energ. Rev. 91 (2018) 109–125, https://doi.org/10.1016/j.rser.2018.03.047.

[2] B. Zakeri, S. Syri, Electrical energy storage systems: a comparative life cycle cost analysis. Renew. Sust. Energ. Rev. 42 (2015) 569–596, https://doi.org/10.1016/j.rser.2014.10.011.

[3] Philips Semiconductor Ltd, NiMH and NiCd battery management. Microprocess. Microsyst. 19 (1995) 165–174, https://doi.org/10.1016/0141-9331(95)90005-5.

[4] S. Ma, M. Jiang, P. Tao, C. Song, J. Wu, J. Wang, T. Deng, W. Shang, Temperature effect and thermal impact in lithium–ion batteries: a review. Prog. Nat. Sci. Mater. Int. 28 (2018) 653–666, https://doi.org/10.1016/j.pnsc.2018.11.002.

[5] B. Zhu, Z. Liang, D. Xia, R. Zou, Metal-organic frameworks and their derivatives for metal-air batteries. Energy Storage Mater. (2019), https://doi.org/10.1016/j.ensm.2019.05.022.

[6] Sodium—sulfur battery development, phase III. J. Power Sources 5 (1980) 356–357, https://doi.org/10.1016/0378-7753(80)80043-7.

[7] C. Verma, E.E. Ebenso, M.A. Quraishi, Corrosion inhibitors for ferrous and non-ferrous metals and alloys in ionic sodium chloride solutions: a review. J. Mol. Liq. 248 (2017) 927–942, https://doi.org/10.1016/j.molliq.2017.10.094.

[8] H.K. Salim, R.A. Stewart, O. Sahin, M. Dudley, Drivers, barriers and enablers to end-of-life management of solar photovoltaic and battery energy storage systems: a systematic literature review. J. Clean. Prod. 211 (2019) 537–554, https://doi.org/10.1016/j.jclepro.2018.11.229.

[9] A. Afif, S.M.H. Rahman, A. Tasfiah Azad, J. Zaini, M.A. Islan, A.K. Azad, Advanced materials and technologies for hybrid supercapacitors for energy storage—a review. J. Energy Storage 25 (2019) 100852, https://doi.org/10.1016/j.est.2019.100852.

[10] J. Ma, Q. Li, M. Kühn, N. Nakaten, Power-to-gas based subsurface energy storage: a review. Renew. Sust. Energ. Rev. 97 (2018) 478–496, https://doi.org/10.1016/j.rser.2018.08.056.

[11] I. Ullah Khan, M.H.D. Othman, H. Hashim, T. Matsuura, A.F. Ismail, M.R.-D. Arzhandi, I.W. Azelee, Biogas as a renewable energy fuel—a review of biogas upgrading, utilisation and storage. Energy Convers. Manag. 150 (2017) 277–294, https://doi.org/10.1016/j.enconman.2017.08.035.

[12] K.-Y. Chan, B. Jia, H. Lin, N. Hameed, J.-H. Lee, K.-T. Lau, A critical review on multifunctional composites as structural capacitors for energy storage. Compos. Struct. 188 (2018) 126–142, https://doi.org/10.1016/j.compstruct.2017.12.072.

[13] P. Mukherjee, V.V. Rao, Design and development of high temperature superconducting magnetic energy storage for power applications—a review. Physica C 563 (2019) 67–73, https://doi.org/10.1016/j.physc.2019.05.001.

[14] G. Alva, Y. Lin, G. Fang, An overview of thermal energy storage systems. Energy 144 (2018) 341–378, https://doi.org/10.1016/j.energy.2017.12.037.

[15] G. Li, Sensible heat thermal storage energy and exergy performance evaluations. Renew. Sust. Energ. Rev. 53 (2016) 897–923, https://doi.org/10.1016/j.rser.2015.09.006.

[16] K.S. Reddy, V. Mudgal, T.K. Mallick, Review of latent heat thermal energy storage for improved material stability and effective load management. J. Energy Storage 15 (2018) 205–227, https://doi.org/10.1016/j.est.2017.11.005.

[17] J. Farfan, C. Breyer, Combining floating solar photovoltaic power plants and hydro-power reservoirs: a virtual battery of great global potential. Energy Procedia 155 (2018) 403–411, https://doi.org/10.1016/j.egypro.2018.11.038.

[18] K.R. Pullen, The status and future of flywheel energy storage. Joule 3 (2019) 1394–1399, https://doi.org/10.1016/j.joule.2019.04.006.

[19] A. Arabkoohsar, L. Machado, R.N.N. Koury, K.A.R. Ismail, Energy consumption minimization in an innovative hybrid power production station by employing PV and evacuated tube collector solar thermal systems. Renew. Energy 93 (2016) 424–441, https://doi.org/10.1016/j.renene.2016.03.003.

[20] A. Arabkoohsar, L. Machado, R.N.N. Koury, Operation analysis of a photovoltaic plant integrated with a compressed air energy storage system and a city gate station. Energy 98 (2016) 78–91, https://doi.org/10.1016/j.energy.2016.01.023.

[21] A. Arabkoohsar, L. Machado, M. Farzaneh-Gord, R.N.N. Koury, Thermo-economic analysis and sizing of a PV plant equipped with a compressed air energy storage system. Renew. Energy 83 (2015), https://doi.org/10.1016/j.renene.2015.05.005.

[22] C.-J. Yang, Chapter 2. Pumped hydroelectric storage. in: T.M. Letcher (Ed.), Storing Energy, Elsevier, Oxford, 2016, pp. 25–38, https://doi.org/10.1016/B978-0-12-803440-8.00002-6.

[23] A. Benato, A. Stoppato, Pumped thermal electricity storage: a technology overview. Therm. Sci. Eng. Prog. 6 (2018) 301–315, https://doi.org/10.1016/j.tsep.2018.01.017.

[24] A. Arabkoohsar, G.B.B. Andresen, Design and analysis of the novel concept of high temperature heat and power storage. Energy 126 (2017) 21–33, https://doi.org/10.1016/j.energy.2017.03.001.

[25] A. Arabkoohsar, G.B. Andresen, Dynamic energy, exergy and market modeling of a high temperature heat and power storage system. Energy 126 (2017), https://doi.org/10.1016/j.energy.2017.03.065.

[26] A. Arabkoohsar, G.B. Andresen, Thermodynamics and economic performance comparison of three high-temperature hot rock cavern based energy storage concepts. Energy 132 (2017), https://doi.org/10.1016/j.energy.2017.05.071.

[27] A. Arabkoohsar, M. Dremark-Larsen, R. Lorentzen, G.B. Andresen, Subcooled compressed air energy storage system for coproduction of heat, cooling and electricity. Appl. Energy 205 (2017) 602–614, https://doi.org/10.1016/j.apenergy.2017.08.006.

[28] A. Arabkoohsar, An integrated subcooled-CAES and absorption chiller system for cogeneration of cold and power. in: 2018 International Conference of Smart Energy System Technology, 2018, pp. 1–5, https://doi.org/10.1109/SEST.2018.8495831.

[29] A. Arabkoohsar, G.B. Andresen, Design and optimization of a novel system for trigeneration. Energy 168 (2019) 247–260, https://doi.org/10.1016/j.energy.2018.11.086.

[30] A.B. Gallo, J.R. Simões-Moreira, H.K.M. Costa, M.M. Santos, E. Moutinho dos Santos, Energy storage in the energy transition context: a technology review. Renew. Sust. Energ. Rev. 65 (2016) 800–822, https://doi.org/10.1016/j.rser.2016.07.028.

CHAPTER TWO

Thermal energy systems

Ahmad Arabkoohsar and Amir Ebrahimi-Moghadam
Department of Energy Technology, Aalborg University, Esbjerg, Denmark

Abstract

This chapter covers thermal energy storage (TES) techniques as a category of mechanical energy storage (MES) methods. In this category of MES, thermal energy (either heat or cold) is stored via the use of a storage medium for a shorter or longer term. TES techniques are categorized into three classes: sensible TES (STES), latent TES (LTES), and thermo-chemical TES (TCTES). In terms of duration of storage, TES systems are classified into short-term and long-term systems. This chapter provides an overview of these techniques as well as mathematical models for each. It also discusses seasonal TES (SeTES), which can be based on either STES or LTES techniques.

2.1 Fundamentals

Due to the recent increase in energy demand, energy storage and the importance of renewable energies have attracted a lot of attention [1]. Besides managing the mismatch between energy demand and supply by conserving and repaying energy, energy storage systems can improve the thermal efficiency and reliability of energy systems [2]. As an energy storage approach, thermal energy storage (TES) is a technology in which thermal energy (either heat or cold) can be stored through a storage medium. At a later time, the stored energy can be used to satisfy the thermal energy needs of the end-user (e.g., a building, a consumer, etc.). As shown in Fig. 2.1, TES systems are categorized into three main groups: sensible TES (STES), latent TES (LTES), and thermo-chemical TES (TCTES) [3].

A comparison of different types of TES technologies is presented in Table 2.1. As shown in the table, STES and TCTES approaches have the lowest and highest energy storage densities, respectively. Also, TCTES systems have the greatest efficiency range. However, the development of TCTES systems is still limited due to their immature R&D background. LTES units are generally smaller than STES units, but LTES units have a higher cost due to costly storage materials [4, 5]. Details related to each of these technologies are presented in the next sections.

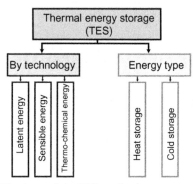

Fig. 2.1 Different classifications of the TES methods.

Table 2.1 Comparison of different TES technologies.

Storage approach	Development level	Efficiency range	Capacity range (kWh/t)	Cost range (€/kWh)
STES	Commercial	50–90	10–50	0.1–10
LTES	R&D/demonstrator/ commercial	75–90	50–150	10–50
TCTES	R&D only	75–100	120–250	8–100

The main costs of TES systems are associated with the costs of the storage materials, technical equipment for charging/discharging, and operation and maintenance. STES systems are rather inexpensive compared to other technologies as they consist of simple reservoirs, cheap storage materials (e.g., water, soil, rocks, etc.), and inexpensive equipment for charging/discharging. Note that although the costs of the storage reservoir are relatively low, it requires effective thermal insulation, which might increase its cost proportionally. Depending on the system size and thermal insulation, STES systems may cost between 0.1 and 10 €/kWh. The costs for an LTES unit with a phase change material (PCM) as the storage medium may be within the range of 10–50 €/kWh, while a TCTES unit costs around 8–100 €/kWh [6].

2.1.1 STES systems

The STES system is considered the simplest type of TES method. In this approach, the thermal energy is stored through heating or cooling a storage medium that is either liquid or solid material (e.g., water, rocks, sand, etc.). The thermophysical properties of the most used heat storage materials in STES

Table 2.2 Thermophysical properties of the most popular storage mediums in STES systems.

Storage medium	Material type	Temperature range (°C)	Specific heat (kJ/kg K)	Density (kg/m³)
Water	Liquid	0–100	4.190	1000
Engine oil	Oil	≤ 160	1.880	888
Calorie HT43	Oil	12–260	2.200	867
Ethanol	Organic liquid	≤ 78	2.400	790
Propane	Organic liquid	≤ 97	2.500	800
Isotunaol	Organic liquid	≤ 100	0.300	808
Butane	Organic liquid	≤ 118	2.400	809
Octane	Organic liquid	≤ 126	2.400	704
Isopentanol	Organic liquid	≤ 148	2.200	831
Sand	Solid	20	0.800	1555
Granite	Solid	20	0.820	2640
Cast iron	Solid	20	0.837	7900
Brick	Solid	20	0.840	1600
Rock	Solid	20	0.879	2560
Concrete	Solid	20	0.880	2240
Aluminum	Solid	20	0.896	2707
NaCl	Solid	200–500	0.850	2160

systems are listed in Table 2.2 [7]. Water is the most popular and commercial storage medium in this category because it is the cheapest option with high specific heat capacity. However, for higher temperatures, heat transfer oils, liquid metals, and so on should be used. Another notable point is that underground STES systems can be used for large-scale applications [8].

In an STES system, the stored/reclaimed thermal energy during a charging/discharging process shows itself in the form of temperature change of the storage medium. Therefore, a higher thermal capacity of the storage medium is very important for having a high energy storage density [9]. Fig. 2.2

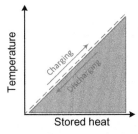

Fig. 2.2 Relation of the stored thermal energy with the storage medium temperature for a STES system.

shows how this happens through charging and discharging processes for sensible heat storage. For cold storage, naturally, the direction of the temperature axis will be reversed.

As mentioned, water is the most common storage medium for STES systems. A STES system with water as the storage medium is mainly a storage tank. In a storage tank, thermal stratification is important for obtaining better charging/discharging performance. Thermal stratification means that water at a higher temperature will have a lower density and thus lies at the upper part of the tank. In contrast, colder water will be denser and lies at the bottom part of the tank. Considering a fully mixed tank is also possible, but this will result in the distribution of the stored energy through all the tank volume and therefore decrease the maximum temperature available in the heat storage tank (or increase the lowest temperature for a cold storage tank) [10].

For a stratified tank, there will be several temperature nodes across the height of the tank. The zone corresponding to each temperature is called a node. Thus, a stratified heat storage tank is a multi-node control volume with inlet and outlet flows from the top and bottom. Fig. 2.3 shows how incoming and outgoing energy flows into and out of the tank affect the energy balance of a multi-node storage tank.

Assuming a storage tank connected to a solar collector set, Fig. 2.4 shows how a stratified tank (left) and a mixed tank (right) will look in terms of

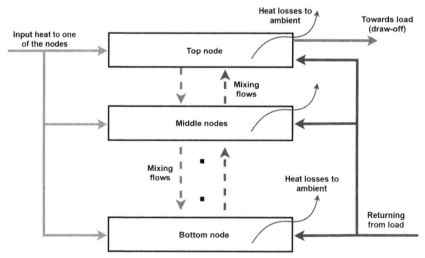

Fig. 2.3 Energy flows in a stratified thermal storage tank [11].

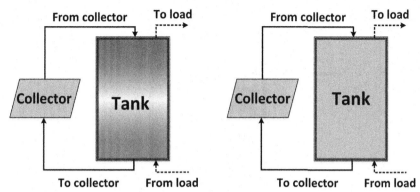

Fig. 2.4 Schematic of stratified tank *(left)*, fully mixed tank *(right)*.

temperature distribution along the height of the tank. The required heat for charging the storage tank is supplied by the solar collectors and then the stored heat is discharged for covering a load for any given needed draw-off. If the tank is well stratified, the highest part of the storage tank will be at a higher temperature and the lowest part will be at a much lower temperature. This causes the tank to be able to supply hot water to the end-user as soon as the top node of the tank only has the required draw-off temperature. On the other hand, a mixed tank, which is at a uniform temperature at all nodes, will not be able to supply any hot water for the end-user until the entire tank reaches the required temperature [12].

Water storage tanks can come in a variety of configurations, including buffer tanks, coiled tanks, and others. In a buffer tank, the incoming hot/cold flows are injected directly into the water bath in the tank and stratification occurs automatically because of the difference in water densities at different temperatures. In a coiled storage tank, which is appropriate for when the working fluid doing the charging is not pure/potable water (or in general, different from the storage medium if not water), the hot working fluid passes through a helix heat exchanger positioned in the storage tank to give its thermal energy to the storage medium. Thermal stratification within the tank is continuously and spontaneously updated as heat is received through the helix at different nodes. Fig. 2.5 depicts the simple schematic of a coiled storage tank.

Just like water storage tanks, the operation principle of other STES methods is also based on temperature difference in a storage medium. For example, Fig. 2.6 shows the schematic of a packed bed of rocks used for high-temperature STES applications. In this system, a working fluid

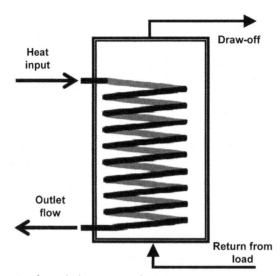

Fig. 2.5 Schematic of a coiled storage tank.

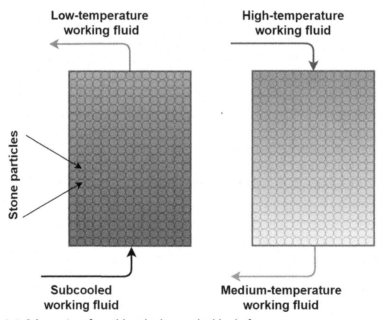

Fig. 2.6 Schematics of a cold and a hot packed bed of stones.

(water, air, etc.) at high (or low) temperature flows through the storage medium (which includes several stone particles). The heat (or cold) energy along the working fluid is transferred to the stones' bed. A packed bed of stones is appropriate for too high or too low temperatures at large scales. Stones are stable at severe temperature and pressure conditions, non-toxic and inflammable, and readily available at very low cost. These features make such storage units quite popular for industrial applications. Employing these for high-temperature heat and power storage systems [13–20]. Fig. 2.6 illustrates cold (left) and hot (right) STES systems in the form of two packed beds of rocks.

2.1.2 LTES systems

In LTES systems, PCMs are the storage mediums and phase change is the mechanism of thermal energy storage. A phase change process takes place at isothermal conditions either from the solid phase to the liquid phase (heat storage) or contrariwise (cold storage) [21]. The energy storage density of LTES systems is much larger than that of STES systems due to large phase change enthalpy of materials compared to that required to make temperature change in them in a given phase. This leads to a much smaller size (volume) of storage if an LTES unit is used instead of a STES system. As an example, Fig. 2.7 shows a schematic of the melting process in an LTES system. As shown, supplying heat to the storage medium increases its temperature when it is in a solid phase. By continuing the heating process, the phase of the storage medium changes from solid to liquid while staying at a constant temperature. As soon as the material is fully liquefied, the temperature increase process starts again. LTES systems usually work in a very narrow range of

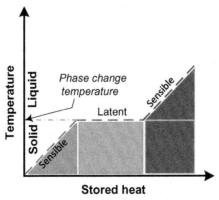

Fig. 2.7 Thermal energy storage process for an LTES system.

temperatures around the phase change point. For example, residential LTES systems usually work in a ±5 K temperature range around the solidification/ liquefaction temperature. This process can be reversed (changing phase from solid to liquid) for cold storage [22].

Today, there is a very broad list of PCMs suitable for use in LTES systems. These can be classified into the three main categories of organic PCMs, inorganic PCMs, and a mixture of other PMCs (so-called eutectics). Conventional PCMs are mainly inorganic, cheap, and present high thermal conductivity factors and latent heat. However, they do have some undesirable side effects such as corrosion. On the other hand, organic PCMs suffer from flammability, low conduction factor, and large volume change while changing phase. However, they are non-corrosive, non-supercoiling, available in nature or via recycling, and available in a large temperature range. The aim of mixed PCMs is to develop new materials that have the advantages of both organic and inorganic PCM classes, but there is still a lack of test data and literature about them. Various PCM types are well discussed in Ref. [22].

The decision about what type of PCM to use depends on the temperature level needed, the type of storage application, economic considerations, specific heat, whether it is organic or inorganic, and some further technical considerations. Table 2.3 presents the melting temperature and latent heat of a number of common PCMs in different categories.

One of the applications of the LTES approach is to cool buildings. Ice TES (ITES) is one of the methods in this category. The storage medium in this technology is crystals or slurries of ice (i.e., solid water). Fig. 2.8 is a schematic sketch of an ice harvesting system. This TES system uses the latent heat of fusion of water. In this system, ice blocks with cold water are used to provide the required cooling load of buildings in fluctuating conditions. Depending on the required cooling load, different ice harvester systems or chillers can be used [25]. In this system, the evaporators are placed over the TES tank where the water return flow is fed into and the ice blocks are formed. Then, these ice blocks are mixed with water in the TES tank.

Ice slurry TES is another type of ITES technology. The main components of an ice slurry system are a slurry ice generator, a compressor/ condenser unit, a TES tank, and a heat exchanger. The slurry ice system is mated to an ice storage tank. The compressor/condenser unit supplies the refrigerant into the ice slurry generator where the pumpable ice slurry is made. The generated ice and water/glycol solution are mixed in the TES tank. Finally, the building's air conditioner unit is linked to the system through the heat exchanger [26]. Fig. 2.9 is an illustration of this technology.

Table 2.3 Thermophysical properties of the most popular PCMs [22–24].

Material	Phase change temperature (°C)	Latent heat (kJ/kg)
Paraffin wax	32–32.1	251
Capric acid	32	152.7
Polyethelene glycol 900 (PEG900)	34	150.5
Lauric-palmitic acid (69:31) eutectic	35.2	166.3
Lauric acid	41–43	211.6
Stearic acid	41–43	211.6
Medicinal paraffin	40–44	146
Paraffin wax	40–53	200–220
P116-Wax	46.7–50	209
Merck P56–58	48.86–58.06	250
Commercial paraffin wax	52.1	243.5
Myristic acid	52.2	182.6
Paraffin RT60/RT58	55 to 60	214.4–232
Palmitic acid	57.8–61.8	185.4
$Mg(NO_3)_2 \cdot 6H_2O$	89	162.8
RT100	99	168
$MgCl_2 \cdot 6H_2O$	116.7	168.6
Erythritol	117.7	339.8
Na/K/NO_3 (0.5/0.5)	220	100.7
$ZnCl_2$/KCl (0.319/0.681)	235	198
$NaNO_3$	310	172
KNO_3	330	266
NaOH	318	165
KOH	380	149.7
$ZnCl_2$	280	75
LiF-CaF_2 (80.5:19.5) mixture	767	816
KOH	380	150
$MgCl_2$/KCl/NaCl	380	400
Na_2CO_3-$BaCO_3$/MgO	500–850	NA
Li_2CO_3	618	NA
Sb_2O_3	652	387
$MgCl_2$	714	542
80.5% LiF–19.5% CaF_2 eutectic	767	790
LiF	850	811
Na_2CO_3	854	276
K_2CO_3	897	236
Water-ice	0	335
GR25	23.2–24.1	45.3
RT25–RT30	26.6	232.0
n-Octadecane	27.7	243.5
$CaCl_2 \cdot 6H_2O$	29.9	187
$Na_2SO_4 \cdot 10H_2O$	32, 39	180

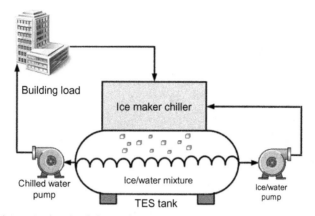

Fig. 2.8 Schematic sketch of the ice harvesting TES.

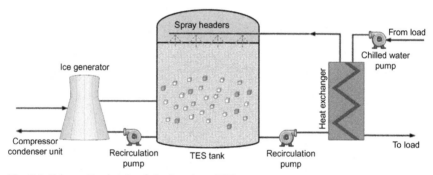

Fig. 2.9 Schematic sketch of the ice slurry TES.

PCM-TES is another type of LTES system. Due to the better thermo-physical properties of PCMs compared to ice, a PCM-TES is a more preferred practical application in buildings. It should be noted that although the energy storage capacity of ITES is greater than that of PCM-TES, the volumetric storage density of PCM-TES is greater compared to ITES. Also, some PCMs (such as organic PCMs) show good thermal stability and proportional freezing and melting features. This is why PCM-TES systems are generally more desirable for application in buildings [27].

2.1.3 TCTES systems

The TCTES method is an emerging approach suitable for high-density and long-term applications. As shown in Fig. 2.10, in this approach the storage medium undergoes a reversible chemical reaction. Firstly, the storage

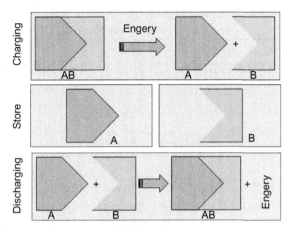

Fig. 2.10 Different processes during TCTES.

medium (i.e., AB) absorbs the energy and then is chemically divided into two components (i.e., A and B). The second step is the storage step, in which the products of the previous step are stored separately until the discharge process. Finally, a reverse reaction occurs at suitable pressure and temperature, and the stored energy is released during this process [28].

As can be seen in Fig. 2.10, the following three processes take place during a roundtrip operation of TCTES systems.

Charging: The charging process is an endothermic reaction. During this process, the storage medium (which is a thermo-chemical material) receives thermal energy. The required energy is clearly the thermal energy to be stored supplied by a renewable source. This energy is supplied for the decomposition of the thermochemical material (i.e., AB) into the two separate materials (i.e., A and B). It is worth mentioning that this energy is equal to the enthalpy of formation for this reaction. Fulfilling the charging process, one has the two separate elements of A and B that should be stored.

Storing: This phase is the storing process of the separated/formed elements of A and B. The process should take place at no or very little thermal energy loss. Thus, storing at ambient temperature is a common storage condition. There are still, of course, some other minor thermal losses that are associated with the destruction of the materials.

Discharging: At a later time, the supplied energy could be recovered again through an exothermic reaction between materials A and B. After this process, component AB is recreated and can be utilized again in the cycle.

Some of the most important thermo-chemical reactions that have been used in TCTES systems are introduced in Table 2.4 [29].

Table 2.4 Some of the thermo-chemical reactions used in TCTES technology.

Reaction	Temperature (K)	Energy density
$CaCO_3 + \Delta H \rightleftharpoons CaO + CO_2$	1169	4400 MJ/m^3
$MnO_2 + \Delta H \rightleftharpoons 1/2M_2O_3 + 1/4O_2$	803	42 kJ/mol
$Ca(OH)_2 + \Delta H \rightleftharpoons CaO + H_2O$	778	3000 MJ/m^3
$NH_3 + \Delta H \rightleftharpoons 1/2N_2 + 3/2H_2$	673–773	67 kJ/mol
$MgH_2 + \Delta H \rightleftharpoons Mg + H_2$	523–773	75 kJ/mol
$MgO + H_2O \rightleftharpoons Mg(OH)_2$	523–673	3300 MJ/m^3
$CH_4 + H_2O \rightleftharpoons CO + 3H_2$	773–1273	–

As the TCTES system is not completely mature, there still are some technical issues that need to be addressed in future research [30, 31]:

- Finding a proper active material or reaction couple that has a high energy density and suitable hydrothermal stability.
- Finding effective factors on different aspects of this technology.
- Better knowledge about different safety issues (e.g., reactants, toxicity, corrosion, etc.).
- More studies about the efficiency of this technology and evaluation of it from an economic point of view.

2.1.4 SeTES systems

SeTES is a way in which the storage of heat or cold is possible for several months. For example, the excess heat of summer could be stored for winter use or, on the contrary, the cooling energy of winter can be stored for summer use. In other words, due to the high demand for space heating in the winter months, there is a mismatch between supply and demand. Thus, one solution is to use the SeTES method, in which energy is collected in the summer, stored seasonally, and used to meet demand in the winter [32]. Hence, the SeTES approach can be considered as an efficient method for smoothening the seasonal energy demand of buildings (or in general heat- and cold-consuming units such as greenhouses) and thereby contributing to mitigate climate change effects and reduce greenhouse gas emissions.

This approach is mainly based on the STES method because of the large storage space and material required, which makes the cost of the system a critical matter. The system cost and performance depend on the mass, specific heat, and temperature change of the storage medium. Hard rocks and water are the most suitable storage mediums for a SeTES system due to their high energy densities and low costs. An SeTES system can come in various designs. Some of the most important kinds are aquifer TES (ATES), bore-hole TES (BTES), and cavern TES (CTES) [33–35].

ATES systems: ATES technology is grouped in the category of under-ground storage systems. Hence, it highly depends on geological conditions of the environment that is going to be utilized. Two of the most important considerations/limitations to be be taken into account in the design of an ATES system are:
- the location of the plant and static head and flow direction of the groundwater
- preventing leakage by installing the ATES system between the impervious layers of the Earth

A schematic of an ATES system and its operation in different operational modes is presented in Fig. 2.11. As shown in the figure, the base required elements in this approach are aquifer wells, extraction pumps, and heat exchangers. During the summer season, the temperature of the groundwater is typically in the range of 5–10°C. Using a pump, cold groundwater is extracted from the aquifer well and supplied into the heat exchanger. Then, the supplied cold can be utilized for any cooling application of the summer season. For cooling a building, for example, the warm air or fluid coming from the building is fed into the heat exchanger. In the heat exchanger, the warm flow rejects its heat into the cold flow (coming from the aquifer well). In this way, the cold demand of the building is supplied. The warmed groundwater returns to the aquifer, leading to the proportional increase of the aquifer bed temperature. It is worth mentioning that depending on the

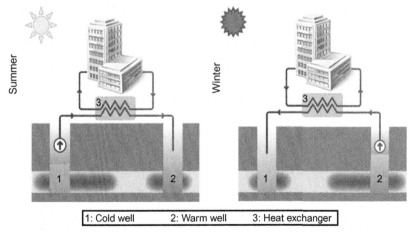

| 1: Cold well | 2: Warm well | 3: Heat exchanger |

Fig. 2.11 Schematic of an ATES system in the winter and summer.

site location and extraction depth, the temperature of the groundwater may be within 5–30°C [36].

A process similar to what happens in the summer can take place in the winter too, but in the reverse way. During the winter, the stored warm groundwater during the summer is fed again into the above-ground heat exchanger, where the supplied heat is to be released into the cold air/fluid coming from the building to provide the heating demand of the building.

ATES systems have the capability of being used for large-scale projects (e.g., up to $15\,kWh/m^3$). In one of the commercial projects in Rostock (Germany), an ATES system was employed to cover the heating load of 108 apartments with a total heating area of $7000\,m^2$. This ATES unit was coupled with a 1000-m^2 solar collector farm that was installed on the buildings' roofs [37].

BTES system: BTES technology is another type of underground TES system similar to the ATES system but with some differences in design configuration. Fig. 2.12 is a schematic view of a BTES system. The main component of this system is U-tube pipe(s) made of synthetic material (generally, high-density polyethylene). Also, as depicted in Fig. 2.13, different configurations of U-tubes used in this technology are introduced (e.g., single tube, double tube, concentric tube-in-tube) [38]. After digging a borehole, this pipe is installed in underground surfaces. The borehole is filled with grout, which is usually a mixture of bentonite and quartz with sand or sand-water. The thermal conductivity of the used grout depends on the components used in the mixture. The heat transfer fluid inside the embedded tube is

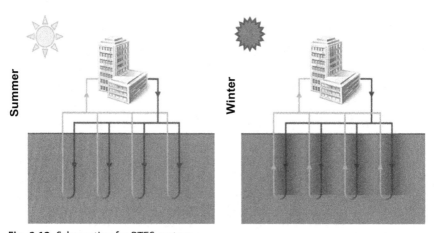

Fig. 2.12 Schematic of a BTES system.

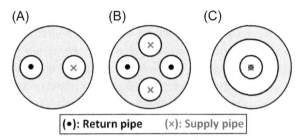

Fig. 2.13 Configuration of (A) single, (B) double, and (C) concentric tube-in-tube U-tubes used in BTES systems.

Table 2.5 Specification of physical properties of typical BTES systems.

Parameter	Range
Borehole diameter/depth	100–150 mm/30–100 m
Distance between boreholes	2–4 m
Minimum/maximum inlet temperature of U-tube pipes	−5°C/>90°C
Average capacity per borehole length	20–30 W/m
Cost of the BTES system per borehole length	50–80 €/m

mainly water in which some glycol or ethanol is added to prevent freezing [39]. The typical values of some specifications of BTES systems are listed in Table 2.5 [40].

Similar to the previously discussed SeTES technologies, there are still some challenges in BTES technology limiting its use for medium- and large-scale applications. These limitations include:

• The high capital costs related to digging deep boreholes are (about 20%–25% of the system's total costs)

• The thermal disturbances/fluctuations that occur in some underground hydrogeological systems

2.1.5 CTES systems

CTES systems are based on creating large water reservoirs in the subsoil for storing heat or cold energies. The hot water CTES and gravel/water CTES are the most well-known approaches in CTES systems. From the technical point of view, these approaches are feasible, but they have high capital costs. The charging/discharging processes are based on the addition/rejection of the energy to/from the storage space by pumping water into/out of the storage system.

Fig. 2.14 shows a schematic of a hot water CTES system. As shown, the storage space is a huge cavern or tank that is isolated carefully and embedded underground. The cavern can be made of rock or be an artificially buried tank [41]. During the charging process, hot water is fed into the cavern from the top, while colder water is extracted from the bottom. When the hot water is injected into the cavern for the first time, the heat losses to the surrounding rock mass are significant [42]. However, after the first year or two, a thermal halo will forms around the cavern with decreasing temperature away from the hot center. Although there will still be some heat losses during the operation of the system even with this thermal halo, the rate of losses will substantially be reduced. One of the most important advantages of hot water CTES systems is their high heat capacity.

Fig. 2.15 illustrates a schematic view of the gravel/water CTES system. In this approach, the storage medium is filled with both water and rocks. Similar to the hot water CTES, at least the sidewalls and top of the underground storage medium of a gravel/water CTES system should be waterproofed and insulated. The storage medium can be waterproofed with a suitable plastic layer and then filled with a mixture of water and gravel. The specific heat capacity of the gravel/water mixture is lower than that of water; hence, bigger storage space should be considered for the gravel/water approach to have the same heat storage capacity in both of the hot water and gravel/water approaches (approximately 50% larger volume is required). The charging/discharging can be done either by direct water exchange or by plastic pipes installed in different layers within the storage medium [43].

It is not unreasonable to consider underground large pit seasonal storage systems in the category of CTES systems. These systems, which work with

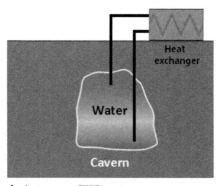

Fig. 2.14 Schematic of a hot water CTES system.

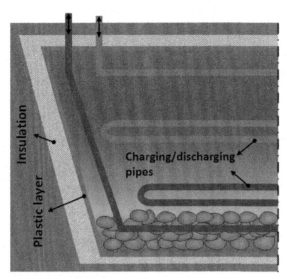

Fig. 2.15 Schematic of a gravel/water CTES system.

water as the main working fluid, include pit-shape underground storage space with the same top level as the ground surface and are connected to a thermal energy supply source and, of course, a consumption source. A good example of such a system is the pit SeTES system recently used for district heating supply in Northern Europe. These SeTES systems are connected to large-scale solar thermal collector farms that harvest solar energy of the summer in the form of hot water, store it in the pit storage, and use it for district heating supply during the winter. Fig. 2.16 presents the schematic of an underground pit SeTES system. As shown in the figure, the pit storage is only insulated thermally from the top as investigations show that insulation is not profitable to be used in all the surrounding areas of a pit SeTES, and it is the top area that critically needs efficient insulation [44]. Large-scale pit SeTES units integrated with solar heating plants and district heating systems are gaining more interest in Europe, especially in Denmark [45]. These plants provide a large portion of the buildings of the country with heat supply for domestic hot water and space heating uses [46]. The Marstal plant (with more than $33,000 \, m^2$ collectors and $75,000 \, m^3$ pit SeTES capacity), the Dronningland plant (with more than $37,000 \, m^2$ collectors and $60,000 \, m^3$ pit SeTES capacity), and the Vojens plant (with $70,000 \, m^2$ collectors and $200,000 \, m^3$ pit SeTES capacity) are only some of the many pit SeTES solar field plants built in Denmark for district heating supply [47].

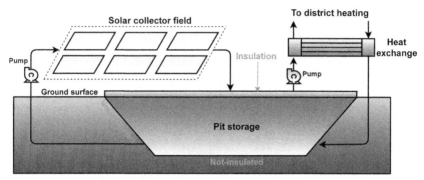

Fig. 2.16 An illustration of a typical pit SeTES system connected to a solar thermal field for district heating supply.

2.2 Mathematical model

The mathematical models related to each of the discussed TES systems are presented in this section.

2.2.1 STES systems

The general form of the energy balance equation for fully mixed (lumped) tanks over a time step can be written as [48]:

$$Q_{st} = \int_{T_i}^{T_f} mc_p dT = mc_p\left(T_f - T_i\right); \Delta T_{st} = T_f - T_i \tag{2.1}$$

$$mc_p\frac{dT_{st}}{dt} = \dot{Q}_{sp} - \dot{Q}_{ld} - \overbrace{U_{st}A_{st}\left(T_i - T_a\right)}^{\dot{Q}_l} \tag{2.2}$$

where, T, m, c_p, and A represent temperature (K), mass (kg), specific heat capacity (J/kgK), and heat transfer area (m^2), respectively. The indices of i, f, and st respectively refer to the initial and final conditions of the time-step and the storage tank. The parameters $\dot{Q}_{sp}$, $\dot{Q}_{ld}$, and $\dot{Q}_l$ are the rate of heat supplied to the tank, the heat withdrawn from the tank for the load, and the rate of heat losses from the tank, respectively. U_S, A_S, and dt represent the coefficient of heat loss from the tank, surface area of the tank, and the time, respectively. Using this equation, one can calculate the rate of the temperature variation of the tank as:

$$T_{st} = T_i + \frac{\Delta t}{mc_p}\left[Q_{sp} - Q_{ld} - U_S A_S(T_i - T_a)\right] \tag{2.3}$$

This is, however, a basic and simple approach. As mentioned, thermal stratification is an important matter in storage tanks. For a stratified tank, a multinode simulation method should be used. This method is based on dividing the storage tank into N nodes and then applying the energy balance equation to each of these nodes. The result is an equation set including N differential equations and N unknowns, which are the temperatures of the N nodes in the next time-step. Assuming a buffer tank, the energy balance equation for each of the nodes can be expressed as follows:

$$m_j c \frac{dT_{st,j}}{dt} = \dot{Q}_{sp} - \dot{Q}_l - \dot{Q}_{ld} + \dot{Q}_{ex} + \dot{Q}_{mix} \tag{2.4}$$

$$m_j c \frac{dT_{st,j}}{dt} = \dot{Q}_{std} \tag{2.5}$$

$$\dot{Q}_{sp} = F_j^{sp} \dot{m}_{sp} c \left(T_{sp} - T_{st,j} \right) \tag{2.6}$$

$$\dot{Q}_l = UA_{l,j} \left(T_{st,j} - T_a \right) \tag{2.7}$$

$$\dot{Q}_{ld} = F_j^{ld} \dot{m}_{ld} c \left(T_{st,j} - T_{ld,r} \right) \tag{2.8}$$

$$\dot{Q}_{ex} = kA_j \frac{\left(T_{st,j-1} + T_{st,j+1} - 2T_{st,j} \right)}{y_j} \tag{2.9}$$

$$\dot{Q}_{mix} = \begin{cases} \dot{m}_{mix,j} c \left(T_{st,j-1} - T_{st,j} \right) & \text{if} \quad \dot{m}_{mix,j} > 0 \\ \dot{m}_{mix,j} c \left(T_{st,j} - T_{st,j+1} \right) & \text{if} \quad \dot{m}_{mix,j+1} < 0 \end{cases} \tag{2.10}$$

where the subscripts and superscripts j, sp, ld, mix, st, l, and r stand for the number of the nodes, hot water flow supplied to the tank, flow rate of load, mixing flows between the nodes, the storage tank, heat losses, and return line from load side, respectively. The heat transfer terms of $\dot{Q}_{std}$, $\dot{Q}_{sp}$, $\dot{Q}_l$, $\dot{Q}_{ld}$, $\dot{Q}_{ex}$, and $\dot{Q}_{mix}$ refer to the rate of stored heat in each node, the injected heat into each node, the lost heat from each node, the heat load flow, the conductive heat transfer between the nodes, and the transferred heat among the nodes due to the mixing flows effect, respectively. Also, the two coefficients of F_j^{sp} and F_j^{ld} respectively show which node the district heating water lies into (F_j^{sp}), and which node the load return flow resides in (F_j^{ld}). These two coefficient are determined as follows:

$$F_j^{sp} = \begin{cases} 1 & if & j=1 \text{ and } T_{sp} > T_{st,j} \\ 1 & if & T_{st,j+1} < T_{sp} < T_{st,j-1} \\ 0 & if & \text{otherwise} \end{cases} \quad (2.11)$$

$$F_j^{ld} = \begin{cases} 1 & if & j=N \text{ and } T_{ld,r} < T_{st,j} \\ 1 & if & T_{st,j+1} < T_{ld,r} < T_{st,j-1} \\ 0 & if & \text{otherwise} \end{cases} \quad (2.12)$$

Based on the preceding equations, during high-temperature mode, $F_1^{sp}=1$ for the first node and $F_j^{sp}=0$ for the other nodes. Also, $F_N^{ld}=1$ for the bottom node and $F_j^{ld}=0$ for the other nodes. The aforementioned governing equations can be solved using various available numerical methods (such as Euler, Crank-Nicolson, Runge-Kutta, etc.).

If the storage tank is a coiled type, the governing equations would be different. Here, the supplied working fluid will not be mixed with the water within the tank and the heat transfer will be in an indirect manner. For this case, again, a multi-node approach should be used and energy balance equations should be developed for all nodes. Thus, one will have [49]:

$$\dot{Q}_{st,j}^{\lambda} = \dot{Q}_{\sup,j}^{\lambda} + \dot{Q}_{ld,j}^{\lambda} + \dot{Q}_{mix,j}^{\lambda} + \dot{Q}_{l,j}^{\lambda} + \dot{Q}_{aux,j}^{\lambda} \quad (2.13)$$

Here, λ is the time-step counter. $\dot{Q}_{st,j}^{\lambda}$ is each node's stored heat in each time-step calculable by:

$$\dot{Q}_{st,j}^{\lambda} = \rho c_p V_j \frac{T_j^{\lambda} - T_j^{\lambda-1}}{\Delta t} \quad (2.14)$$

In which V_j is the volume of water of each node.

Similar to that presented for a buffer tank, the heat load withdrawn from each node can be calculated by:

$$\dot{Q}_{ld,j}^{\lambda} = \dot{m}_{ld,j}^{\lambda} c_p \left(T_j^{\lambda} - T_{j+1}^{\lambda} \right) \quad (2.15)$$

Where

$$\dot{m}_{ld,j}^{\lambda} = \frac{\dot{Q}_{ld}^{\lambda}}{c_p \left(T_{st,1}^{\lambda} - T_{ld,r}^{\lambda} \right)} \quad (2.16)$$

Having Δy as the distance of the central points of the neighboring nodes and k as the thermal conductivity of water, the rate of exchanged heat with the neighboring nodes will be calculated by:

$$\dot{Q}^{\lambda}_{mix,j} = \frac{k}{\Delta y} A_{j+1} \left(T^{\lambda}_j - T^{\lambda}_{j+1} \right) + \frac{k}{\Delta y} A_{j-1} \left(T^{\lambda}_j - T^{\lambda}_{j-1} \right) \qquad (2.17)$$

For the rate of heat losses from each node, the same correlation as Eq. (2.7) is used.

$\dot{Q}^{\lambda}_{\text{sup},j}$ is the rate heat provided by each section of the helix across each node of the tank. Usually, the number of nodes chosen for the helix should be greater than that of the tank to increase the accuracy of calculations. Thus, dividing the helix into m segments, the energy balance of helix segments (nodes) will be as:

$$\rho c_p V_m \frac{\overline{T}^{\lambda}_m - \overline{T}^{\lambda-1}_m}{\Delta t} + \dot{m}^{\lambda}_m c_p \left(T^{\lambda}_{m+1} - T^{\lambda}_m \right) + UA^{\lambda}_m \left(\overline{T}^{\lambda}_m - T^{\lambda}_{m,\,\text{ext}} \right)$$
$$= 0; \text{ where}: \overline{T}^{\lambda}_m = \frac{\left(T^{\lambda}_{m+1} + T^{\lambda}_m \right)}{2} \qquad (2.18)$$

where V_m, $\dot{m}^{\lambda}_m$, T^{λ}_m, and T^{λ}_{m+1} are the segments' volume, the mass flow rate through each segment, and the temperature of working fluid at the entrance and exit of each segment of the helix, respectively. Also, $T^{\lambda}_{m,\,ext}$ is the helix segment temperature at the surface within the tank. Finally, the rate of heat supplied by the helix to each node of the tank is given by:

$$\dot{Q}^{\lambda}_{\text{sup},j} = \sum_{m=r}^{R} UA^{\lambda}_m \left(\overline{T}^{\lambda}_m - T^{\lambda}_{m,\,\text{ext}} \right); \text{ where}: T^{\lambda}_{m,\,\text{ext}} = T^{\lambda}_j \qquad (2.19)$$

Here again numerical methods are required to solve the preceding complex set of equations [50].

2.2.2 LTES systems

Fundamentals of LTES systems were presented in previous sections. Taking into account Fig. 2.7, the amount of stored heat in a PCM (as the storage medium) can be calculated through the summation of the sensible and latent heats [48]:

$$Q_{st} = \int_{T_i}^{T_m} m c_p dT + (m\Delta H) + \int_{T_m}^{T_f} m c_p dT \qquad (2.20)$$

This equation can be rewritten as:

$$Q_{st} = m \left[c_{p,s}(T_m - T_i) + \Delta H + c_{p,l} \left(T_f - T_m \right) \right] \qquad (2.21)$$

$$Q_{st} = m\left[c_{p,s}\Delta T_s + \Delta H + c_{p,l}\Delta T_l\right]; \Delta T_s = T_m - T_i, \Delta T_l = T_f - T_m \quad (2.22)$$

Stephan condition (also known as moving boundary condition) is the main difficulty of solving the heat transfer equations in solid–liquid phase-change processes. The inherent nature of such processes is non-linear and this makes the prediction of the behavior of PCMs difficult. Taking into account the available literature around the mathematical formulation of using PCMs in LTES systems, various formulations can be found. In one study, Lacroix [51] introduced an LTES system based on the use of PCM in a shell-and-tube. The schematic of his proposed model is depicted in Fig. 2.17.

As shown, the model is based on a shell-and-tube unit in which the shell side is filled with the PCM and the heat transfer fluid (HTF) flows through the tube. The energy conservation equation for the PCM and HTF is as follows:

For the PCM:

$$\frac{\partial h}{\partial t} = \frac{1}{r}\frac{\partial}{\partial r}\left(\propto r\frac{\partial h}{\partial r}\right) + \frac{\partial}{\partial z}\left(\propto \frac{\partial h}{\partial z}\right) - \rho\Delta h_f\frac{\partial f}{\partial t} \quad (2.23)$$

For the HTF:

$$\left(\rho c_p\right)_{HTF}\pi R_i^2\frac{\partial T_{HTF}}{\partial t} = 2\pi R_i U(T - T_{HTF}) - \left(\dot{m}c_p\right)_{HTF}\frac{\partial T_{HTF}}{\partial z} \quad (2.24)$$

in which h and $\propto$ are the specific enthalpy and thermal diffusivity of the PCM, Δh_f is the latent heat of fusion, and f is the liquid fraction of the melt. T and T_{HTF} refer to the temperatures of the PCM and HTF, respectively.

Based on the principles discussed for Fig. 2.7 and Crank's equation [52], the enthalpy terms have a relation as Eq. (2.25). This equation states that in the phase-change problems, the total enthalpy is composed of sensible and latent terms.

$$H(T) = h(T) + \rho_s f\Delta h_f \quad (2.25)$$

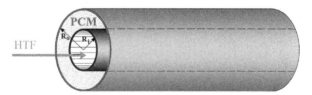

Fig. 2.17 Schematic of an LTES unit used in [51].

$$h(T) = \int_{T_m}^{T} \rho_k c_{p,k} dT \tag{2.26}$$

in which the index k refers to each phase in the PCM and T_m is the phase change temperature.

In another work, Morrison and Abdel-Khalik [53] used a PCM TES unit for energy storage purposes in a solar heating system. They used the following energy balance equations:

For the PCM:

$$\frac{\partial u}{\partial t} = \frac{k}{\rho} \frac{\partial^2 T}{\partial z^2} + \frac{UP}{\rho A} (T_{HTF} - T) \tag{2.27}$$

For the HTF:

$$\frac{\partial T_{HTF}}{\partial t} + \frac{\dot{m}}{\rho_{HTF} A_{HTF}} \frac{\partial T_{HTF}}{\partial z} = \frac{UP}{(\rho c_p)_{HTF} A_{HTF}} (T - T_{HTF}) \tag{2.28}$$

in which u and A are the specific internal energy and area, respectively.

2.2.3 TCTES systems

As mentioned in the previous sections, TCTES systems are based on the decomposition of a thermochemical material (AB). The charging reaction is expressed as:

$$AB + Heat \rightarrow A + B \tag{2.29}$$

The discharging reaction can be written as:

$$A + B \rightarrow AB + Heat \tag{2.30}$$

As an example, the charging and discharging processes for $SrBr_2 \cdot 6H_2O$ can be written as the following equations. During these processes, the reactive element (H_2O), $SrBr_2 \cdot 6H_2O$, and $SrBr_2 \cdot H_2O$ remain in the vapor phase, solid phase, and solid phase, respectively.

$$SrBr_2 \cdot 6H_2O + Heat \rightarrow SrBr_2 \cdot H_2O + 5H_2O, Charging\ reaction \tag{2.31}$$

$$SrBr_2 \cdot H_2O + 5H_2O \rightarrow SrBr_2 \cdot 6H_2O + Heat, Discharging\ reaction \tag{2.32}$$

2.2.4 SeTES systems

The formulation of the underground SeTES systems is composed of two parts of the flow in the ground and flows within pipes. For this, first, one should have the primary knowledge of the governing equation on the fluid motion through the aforementioned control volumes. Then, heat transfer formulations can be presented.

2.2.4.1 Hydraulics of groundwater

The mass conservation equation (known as continuity equation) is the first fundamental equation. If the groundwater flows through a control volume of a porous media type, considering Cartesian coordinates, the continuity equation can be expressed as:

$$\frac{\partial(\rho v_x)}{\partial x} + \frac{\partial(\rho v_y)}{\partial y} + \frac{\partial(\rho v_z)}{\partial z} = -\frac{\partial(\rho n)}{\partial t} \tag{2.33}$$

where ρ is the density and v_x, v_y, and v_z stand for the components of the velocity vector (V_i) in directions of x, y, and z, respectively. Also, the right-hand term in the preceding equation represents the time-rate of changing the stored mass in the control volume and is expressed as follows:

$$\frac{\partial(\rho n)}{\partial t} = \rho S_s \frac{\partial h}{\partial t} \tag{2.34}$$

In underground SeTES systems, the flow of the groundwater generally follows Darcy's law. The average Darcy velocity (u) is described as:

$$u = -K\nabla h, v = \frac{u}{n} \tag{2.35}$$

By applying Darcy's law into the continuity equation, the equation of the hydraulic head distribution within the porous media is derived as:

$$\frac{\partial}{\partial x_i}\left(K_{ij}\frac{\partial h}{\partial x_i}\right) + R = S_s \frac{\partial h}{\partial t} \tag{2.36}$$

$$S_s = \rho g(\alpha + n\beta) \tag{2.37}$$

where K_{ij}, h, R, g, α, and β are the ground's hydraulic conductivity tensor (m/s), hydraulic head (m), source/sink (1/s), acceleration of gravity (m/s^2), the compressibility of the aquifer (1/N/m^2), and the compressibility of water (1/N/m^2), respectively.

The preceding equations can be used in most cases of underground SeTES systems. But in some cases (e.g., in some situations of the fluid flow in BTES systems), Darcy flow can't be considered. In such situations, the relation of the flux and gradient is not linear. But by applying the effective hydraulic conductivity of K_{eff} instead of K_{ij} in Eq. (2.35), it can still be used in BTES systems.

$$K_{eff} = \begin{cases} K \text{ Darcy flow in ATESs} \\ \dfrac{d^2 \rho_l g}{32 \mu_l} \text{ Laminar flow in BTESs} \\ \dfrac{2gd}{uf} \text{ Transitional and turbulent flow in BTESs} \end{cases} \quad (2.38)$$

in which d, μ, and f stand for tube diameter (m), dynamic viscosity (Pas), and friction factor, respectively. Also, the indices l and s are respectively related to the liquid and solid phases.

2.2.4.2 Hydraulics of pipes

The knowledge about pressure drop of fluid flows inside the pipe and the flow regime are important issues in SeTES systems (the pressure drop is especially important in BTES systems for sizing of the circulation pump). The pressure drop within the borehole heat exchangers can be estimated using the following equation [54]:

$$\Delta p = \frac{f l \rho_f U^2}{d} \frac{U^2}{2} + \sum_{i=1}^{n} \epsilon_i \frac{\rho_f U^2}{2} \quad (2.39)$$

in which f and ϵ are the friction factors of the borehole and pipe fixtures, respectively. Also, ρ_f and U show the density (kg/m^3) and velocity (m/s) of the fluid flowing inside the pipe, and d is the diameter of the pipe (m). The borehole friction factor depending on the flow regime is either laminar or turbulent. The Reynolds number (Re) is used for figuring out the flow regime. For internal pipe flow, Re is defined as follows:

$$Re = \frac{\rho_f U d}{\mu_f} = \frac{U d}{\vartheta_f} \quad (2.40)$$

in which μ_f and ϑ_f are respectively the dynamic and kinematic viscosity of the fluid.

Depending on the amount of Re, the following conditions can be drawn for the fluid flow inside a pipe:

- Re<2300: Laminar flow
- $2300 \leq \text{Re} \leq 10^4$: Transient flow between laminar and turbulent flows
- $\text{Re} > 10^4$: Turbulent flow

Finally, the following correlations can be used to calculate the friction factor [55]:

$$f = \begin{cases} \dfrac{64}{\text{Re}}, \text{Laminar} \\ \dfrac{0.3164}{\text{Re}^{\frac{1}{4}}}, \text{Turbulent} \end{cases} \qquad (2.41)$$

2.2.4.3 Governing equations of the heat transfer

After knowing the governing equations of the fluids flows, the heat transfer mechanism can be formulated. The heat transfer mechanism in an underground SeTES is composed of different heat transfer processes:

- Conductive heat transfer through the fluid and phases
- Convective heat transfer from the fluid phase to the solid phase
- Convective heat transfer through the fluid phase (advection)

2.2.4.4 Heat transfer in groundwater

Here, a two-phase heat transfer is considered. The fluid phase represents the groundwater flowing within pores and the solid phase refers to the rocks. The energy balance equation for these phases is expressed as follows:

$$\left(\rho c_p\right)_l \frac{\partial T_l}{\partial t} + \left(\rho c_p\right)_l V_l \nabla T_l = \nabla.(k_l \nabla T_l) + h\frac{A}{V}(T_s - T_l), \text{Fluid} - \text{phase} \quad (2.42)$$

$$\left(\rho c_p\right)_s \frac{\partial T_s}{\partial t} = \nabla.(k_s \nabla T_s) + h\frac{A}{V}(T_l - T_s), \text{Solid} - \text{phase} \qquad (2.43)$$

in which the indies l and s respectively refer to the liquid (i.e., groundwater) and solid (i.e., rock) phases. Also, the parameters T, c_p, t, k, V_l, and $\frac{A}{V}$ are temperature (K), specific heat capacity (J/kg K), time (s), thermal conductivity (W/m K), the velocity vector of groundwater (m/s), and heat transfer area of A in a volume of V (1/m).

It is worth mentioning that in saturated water conditions, due to the thermal equilibrium assumption, it is assumed that the fluid and solid phases are at the same temperature. Accordingly, the average temperature can be considered for both phases. The following equation can be used for the energy balance of each phase:

$$\left(\rho c_p\right)_{eff} \frac{\partial T}{\partial t} = -\left(\rho c_p\right)_l n V_l \nabla \cdot \left(k_{eff} \nabla T\right) \tag{2.44}$$

The following equation can be used for calculating the effective volumetric heat capacity based on the volumetric heat capacity of each phase:

$$\left(\rho c_p\right)_{eff} = n\left(\rho c_p\right)_l + (1-n)\left(\rho c_p\right)_s \tag{2.45}$$

The calculation of the effective thermal conductivity is more complex. The following equation (geometric mean) can be used for effective thermal conductivity. However, an arithmetic formula (such as that used in Eq. 2.44) can also be used for effective thermal conductivity.

$$k_{eff} = k_l^n k_s^{(1-n)} \tag{2.46}$$

Also, the ratio of the thermal conductivity and volumetric heat capacity known as the thermal diffusivity is:

$$\propto_{eff} = \frac{k_{eff}}{\left(\rho c_p\right)_{eff}} \tag{2.47}$$

2.2.4.5 Heat transfer in pipes

There are two heat transfer processes for the flow in the pipes. One is the heat transfer inside the pipe. The thermal conductivity of the fluids in the SeTES is low and hence the axial conductive heat transfer can be ignored in comparison to the convection. Accordingly, the energy balance equation for the fluid flowing inside the pipe can be written as Eq. (2.48). The second is the heat transfer process through the pipe wall. This heat is released into the fluid from the surroundings and is calculated using Eq. (2.49).

$$\left(\rho c_p\right)_f \frac{\partial T_f}{\partial t} \approx -\left(\rho c_p\right)_f V_f \nabla T_f \tag{2.48}$$

$$\left(\rho c_p\right)_P \frac{\partial T_P}{\partial t} = \nabla \cdot (k_P \nabla T_P) + h \frac{A}{V}\left(T_f - T_P\right) \tag{2.49}$$

where the indices f and P refer to the fluid flowing inside the pipe and pipe wall, respectively.

2.3 Future perspective

Today, the use of thermal energy systems is quite common due to their high efficiency and low cost. However, TES systems will be of much greater importance in the future when more and more renewable energy generation systems are implemented and different energy sectors/systems are to be integrated for higher efficiency and lower cost of production and distribution. Thus, TES systems, together with other energy storage technologies including electricity storage systems, will be the key elements of any energy matrix in the future. The main types of TES methods are STES, LTES, and TCTES. Among these methods, STES and LTES approaches are more conventional and quite mature, while TCES techniques are still under development.

TES systems, especially, will be widely used in future district heating and cooling networks. Most of the practical projects in this framework are related to the use of STES and SeTES systems in district energy systems. Using SeTES units coupled with solar energy fields as well as centralized and decentralized smaller and large-scale heat and cold storage units in district heating and cooling systems is already quite common. Long-term TES systems provide a good possibility to reach high-efficiency utilization of renewable energy resources (solar and geothermal) and waste heats for district energy systems. From an economic point of view, the costs associated with TES systems depend on the scale and number of storage cycles. Hence, although STES systems have lower investment costs, they are not suitable for high capacities and high-temperature applications. LTES systems, due to being available in a wide range of temperature levels and having very high storage density compared to that of STES systems, have received special attention in recent decades. For the LTES system, there is still a need for more research on the performance of inorganic PCMs to reach high-efficiency TES systems. Regarding TCTES techniques, despite their high costs, they are considered as appropriate solutions for high-capacity and high-temperature applications. TCTES systems, however, have the lowest storage period among all TES systems (hours to days). As TCTES units could be replaced with some high-cost/low-density electrochemical batteries, this TES method has increasing importance in the future of energy storage.

References

[1] A. Arabkoohsar, Dynamic Modeling of a Compressed Air Energy Storage System in a Grid Connected Photovoltaic Plant, PhD Thesis. (2016).

[2] A. Arabkoohsar, G.B. Andresen, Design and optimization of a novel system for trigeneration. Energy 168 (2019) 247–260, https://doi.org/10.1016/j.energy.2018.11.086.

[3] G. Alva, Y. Lin, G. Fang, An overview of thermal energy storage systems. Energy 144 (2018) 341–378, https://doi.org/10.1016/j.energy.2017.12.037.

[4] D. Lefebvre, F.H. Tezel, A review of energy storage technologies with a focus on adsorption thermal energy storage processes for heating applications. Renew. Sust. Energ. Rev. 67 (2017) 116–125, https://doi.org/10.1016/j.rser.2016.08.019.

[5] E. Guelpa, V. Verda, Thermal energy storage in district heating and cooling systems: a review. Appl. Energy 252 (2019) 113474, https://doi.org/10.1016/j.apenergy.2019.113474.

[6] A. Hauer, Thermal Energy Storage: Technology Brief, IRENA, 2013.

[7] I. Sarbu, C. Sebarchievici, A comprehensive review of thermal energy storage. Sustainability 10 (2018), https://doi.org/10.3390/su10010191.

[8] S. Ayyappan, K. Mayilsamy, V.V. Sreenarayanan, Performance improvement studies in a solar greenhouse drier using sensible heat storage materials. Heat Mass Transf. 52 (2016) 459–467, https://doi.org/10.1007/s00231-015-1568-5.

[9] K.S. Reddy, V. Mudgal, T.K. Mallick, Review of latent heat thermal energy storage for improved material stability and effective load management. J. Energy Storage 15 (2018) 205–227, https://doi.org/10.1016/j.est.2017.11.005.

[10] A. Arabkoohsar, L. Machado, R.N.N. Koury, K.A.R. Ismail, Energy consumption minimization in an innovative hybrid power production station by employing PV and evacuated tube collector solar thermal systems. Renew. Energy 93 (2016) 424–441, https://doi.org/10.1016/j.renene.2016.03.003.

[11] A. Moallemi, A. Arabkoohsar, F.J.P. Pujatti, R.M. Valle, K.A.R. Ismail, Non-uniform temperature district heating system with decentralized heat storage units, a reliable solution for heat supply. Energy 167 (2018) 80–91, https://doi.org/10.1016/j.energy.2018.10.188.

[12] P.H. Feng, B.C. Zhao, R.Z. Wang, Thermophysical heat storage for cooling, heating, and power generation: a review. Appl. Therm. Eng. 166 (2020) 114728, https://doi.org/10.1016/j.applthermaleng.2019.114728.

[13] A. Arabkoohsar, G.B. Andresen, Thermodynamics and economic performance comparison of three high-temperature hot rock cavern based energy storage concepts. Energy 132 (2017), https://doi.org/10.1016/j.energy.2017.05.071.

[14] A. Arabkoohsar, G.B. Andresen, Dynamic energy, exergy and market modeling of a high temperature heat and power storage system. Energy 126 (2017), https://doi.org/10.1016/j.energy.2017.03.065.

[15] A. Arabkoohsar, Combined steam based high-temperature heat and power storage with an organic Rankine cycle, an efficient mechanical electricity storage technology. J. Clean. Prod. 119098 (2019), https://doi.org/10.1016/j.jclepro.2019.119098.

[16] A. Arabkoohsar, Combination of air-based high-temperature heat and power storage system with an organic Rankine cycle for an improved electricity efficiency. Appl. Therm. Eng. 167 (2020) 114762, https://doi.org/10.1016/j.applthermaleng.2019.114762.

[17] W.K. Hussam, H. Rahbari, A. Arabkoohsar, Off-design operation analysis of air-based high-temperature heat and power storage, Energy 196 (2019) 117149.

[18] Siemens, Siemens High Temeprature Heat and Power Storage Project, https://www.siemens.com/press/en/pressrelease/?press=/en/pressrelease/2016/windpower-renewables/pr2016090419wpen.htm&content=WP, 2016.

[19] A. Benato, Performance and cost evaluation of an innovative pumped thermal electricity storage power system. Energy 138 (2017) 419–436, https://doi.org/10.1016/j.energy.2017.07.066.

[20] A. Benato, A. Stoppato, Pumped thermal electricity storage: a technology overview. Therm. Sci. Eng. Prog. 6 (2018) 301–315, https://doi.org/10.1016/j.tsep.2018.01.017.

[21] M.M. Farid, A.M. Khudhair, S.A.K. Razack, S. Al-Hallaj, A review on phase change energy storage: materials and applications. Energy Convers. Manag. 45 (2004) 1597–1615, https://doi.org/10.1016/j.enconman.2003.09.015.

[22] H. Zhang, J. Baeyens, G. Cáceres, J. Degrève, Y. Lv, Thermal energy storage: recent developments and practical aspects. Prog. Energy Combust. Sci. 53 (2016) 1–40, https://doi.org/10.1016/j.pecs.2015.10.003.

[23] A. Nematpour Keshteli, M. Sheikholeslami, Nanoparticle enhanced PCM applications for intensification of thermal performance in building: a review. J. Mol. Liq. 274 (2019) 516–533, https://doi.org/10.1016/j.molliq.2018.10.151.

[24] F. Agyenim, N. Hewitt, P. Eames, M. Smyth, A review of materials, heat transfer and phase change problem formulation for latent heat thermal energy storage systems (LHTESS). Renew. Sust. Energ. Rev. 14 (2010) 615–628, https://doi.org/10.1016/j.rser.2009.10.015.

[25] S. Kalaiselvam, R. Parameshwaran, S. Kalaiselvam, R. Parameshwaran, Chapter 5. Latent thermal energy storage. in: Thermal Energy Storage Technologies for Sustainability, Academic Press, Boston, 2014, pp. 83–126, https://doi.org/10.1016/B978-0-12-417291-3.00005-0.

[26] Y.H. Yau, B. Rismanchi, A review on cool thermal storage technologies and operating strategies. Renew. Sust. Energ. Rev. 16 (2012) 787–797, https://doi.org/10.1016/j.rser.2011.09.004.

[27] J.M. Mahdi, S. Lohrasbi, E.C. Nsofor, Hybrid heat transfer enhancement for latent-heat thermal energy storage systems: a review. Int. J. Heat Mass Transf. 137 (2019) 630–649, https://doi.org/10.1016/j.ijheatmasstransfer.2019.03.111.

[28] R.-J. Clark, A. Mehrabadi, M. Farid, State of the art on salt hydrate thermochemical energy storage systems for use in building applications. J. Energy Storage 27 (2020) 101145, https://doi.org/10.1016/j.est.2019.101145.

[29] S. Kuravi, Y. Goswami, E.K. Stefanakos, M. Ram, C. Jotshi, S. Pendyala, et al., Thermal energy storage for concentrating solar power plants. Technol Innov 14 (2012) 81–91, https://doi.org/10.3727/194982412X13462021397570.

[30] D. Aydin, S.P. Casey, S. Riffat, The latest advancements on thermochemical heat storage systems. Renew. Sust. Energ. Rev. 41 (2015) 356–367, https://doi.org/10.1016/j.rser.2014.08.054.

[31] L. Scapino, H.A. Zondag, J. Van Bael, J. Diriken, C.C.M. Rindt, Sorption heat storage for long-term low-temperature applications: a review on the advancements at material and prototype scale. Appl. Energy 190 (2017) 920–948, https://doi.org/10.1016/j.apenergy.2016.12.148.

[32] R. McKenna, D. Fehrenbach, E. Merkel, The role of seasonal thermal energy storage in increasing renewable heating shares: a techno-economic analysis for a typical residential district. Energy Buildings 187 (2019) 38–49, https://doi.org/10.1016/j.enbuild.2019.01.044.

[33] V. Rostampour, M. Jaxa-Rozen, M. Bloemendal, J. Kwakkel, T. Keviczky, Aquifer thermal energy storage (ATES) smart grids: large-scale seasonal energy storage as a distributed energy management solution. Appl. Energy 242 (2019) 624–639, https://doi.org/10.1016/j.apenergy.2019.03.110.

[34] A. Dahash, F. Ochs, M.B. Janetti, W. Streicher, Advances in seasonal thermal energy storage for solar district heating applications: a critical review on large-scale hot-water

tank and pit thermal energy storage systems. Appl. Energy 239 (2019) 296–315, https://doi.org/10.1016/j.apenergy.2019.01.189.

[35] S.K. Shah, L. Aye, B. Rismanchi, Seasonal thermal energy storage system for cold climate zones: a review of recent developments. Renew. Sust. Energ. Rev. 97 (2018) 38–49, https://doi.org/10.1016/j.rser.2018.08.025.

[36] M. Pellegrini, M. Bloemendal, N. Hoekstra, G. Spaak, A. Andreu Gallego, J. Rodriguez Comins, et al., Low carbon heating and cooling by combining various technologies with aquifer thermal energy storage. Sci. Total Environ. 665 (2019) 1–10, https://doi.org/10.1016/j.scitotenv.2019.01.135.

[37] T. Schmidt, D. Mangold, H. Müller-Steinhagen, Central solar heating plants with seasonal storage in Germany. Sol. Energy 76 (2004) 165–174, https://doi.org/10.1016/j.solener.2003.07.025.

[38] H.-J.G. Diersch, D. Bauer, 7—analysis, modeling and simulation of underground thermal energy storage (UTES) systems. in: L. Cabeza (Ed.), Advances in Thermal Energy Storage System: Methods and Applications, Woodhead Publishing, 2015, pp. 149–183, https://doi.org/10.1533/9781782420965.1.149.

[39] G. Emmi, A. Zarrella, M. De Carli, A. Galgaro, An analysis of solar assisted ground source heat pumps in cold climates. Energy Convers. Manag. 106 (2015) 660–675, https://doi.org/10.1016/j.enconman.2015.10.016.

[40] S. Kalaiselvam, R. Parameshwaran, S. Kalaiselvam, R. Parameshwaran (Eds.), Chapter 7. Seasonal thermal energy storage, Academic Press, Boston, 2014, pp. 145–162, https://doi.org/10.1016/B978-0-12-417291-3.00007-4.

[41] C.R. Matos, J.F. Carneiro, P.P. Silva, Overview of large-scale underground energy storage technologies for Integration of renewable energies and criteria for reservoir identification. J. Energy Storage 21 (2019) 241–258, https://doi.org/10.1016/j.est.2018.11.023.

[42] L. Miró, J. Gasia, L.F. Cabeza, Thermal energy storage (TES) for industrial waste heat (IWH) recovery: a review. Appl. Energy 179 (2016) 284–301, https://doi.org/10.1016/j.apenergy.2016.06.147.

[43] G. Li, X. Zheng, Thermal energy storage system integration forms for a sustainable future. Renew. Sust. Energ. Rev. 62 (2016) 736–757, https://doi.org/10.1016/j.rser.2016.04.076.

[44] T. Schmidt, P. Alex, T. Schmidt, R. Djebbar, R. Boulter, J. Thornton, et al., Design aspects for large-scale pit and aquifer thermal energy storage for district heating and cooling. Energy Procedia 149 (2018) 585–594, https://doi.org/10.1016/j.egypro.2018.08.223.

[45] Z. Tian, B. Perers, S. Furbo, J. Fan, Thermo-economic optimization of a hybrid solar district heating plant with flat plate collectors and parabolic trough collectors in series. Energy Convers. Manag. 165 (2018) 92–101, https://doi.org/10.1016/j.enconman.2018.03.034.

[46] A. Arabkoohsar, A.S. Alsagri, A new generation of district heating system with neighborhood-scale heat pumps and advanced pipes, a solution for future renewable-based energy systems. Energy 193 (2020) 116781, https://doi.org/10.1016/j.energy.2019.116781.

[47] J. Fan, J. Huang, O.L. Andersen, S. Furbo, Thermal performance analysis of a solar heating plant. in: ISES Solar World Conference 2017-IEA SHC International Conference of Solar Heating and Cooling for Buildings and Industry 2017, Proceedings 2017, 2017, pp. 291–300, https://doi.org/10.18086/swc.2017.06.05.

[48] A. Kumar, S.K. Shukla, A review on thermal energy storage unit for solar thermal power plant application. Energy Procedia 74 (2015) 462–469, https://doi.org/10.1016/j.egypro.2015.07.728.

[49] J. Cadafalch, D. Carbonell, R. Consul, R. Ruiz, Modelling of storage tanks with immersed heat exchangers. Sol. Energy 112 (2015) 154–162, https://doi.org/10.1016/j.solener.2014.11.032.

[50] A. Arabkoohsar, Non-uniform temperature district heating system with decentralized heat pumps and standalone storage tanks. Energy 170 (2019) 931–941, https://doi.org/10.1016/j.energy.2018.12.209.

[51] M. Lacroix, Numerical simulation of a shell-and-tube latent heat thermal energy storage unit. Sol. Energy 50 (1993) 357–367, https://doi.org/10.1016/0038-092X(93)90029-N.

[52] J. Crank, Free and Moving Boundary Problems, Oxford University Press, 1987.

[53] D.J. Morrison, S.I. Abdel-Khalik, Effects of phase-change energy storage on the performance of air-based and liquid-based solar heating systems. Sol. Energy 20 (1978) 57–67, https://doi.org/10.1016/0038-092X(78)90141-X.

[54] M. Khosravi, A. Arabkoohsar, Thermal-hydraulic performance analysis of twin-pipes for various future district heating schemes. Energies 12 (2019) 1299, https://doi.org/10.3390/en12071299.

[55] A. Arabkoohsar, M. Khosravi, A.S. Alsagri, CFD analysis of triple-pipes for a district heating system with two simultaneous supply temperatures. Int. J. Heat Mass Transf. 141 (2019) 432–443, https://doi.org/10.1016/j.ijheatmasstransfer.2019.06.101.

CHAPTER THREE

Compressed air energy storage system

Ahmad Arabkoohsar

Department of Energy Technology, Aalborg University, Esbjerg, Denmark

Abstract

A compressed air energy storage (CAES) system is an electricity storage technology under the category of mechanical energy storage (MES) systems, and is most appropriate for large-scale use and longer storage applications. In a CAES system, the surplus electricity to be stored is used to produce compressed air at high pressures. When there is a need for producing extra power, the stored compressed air is used to drive air expanders and thereby actuate an electricity generator. A CAES system may come in a variety of configurations. Each of these designs is appropriate for specific operating conditions, offers its own advantages, and may suffer from its own disadvantages. This chapter gives a fundamental understanding of CAES technology in various configurations followed by a history of the development of this technology over the years in practice and in the literature. Then, it presents the required mathematical model for analyzing a CAES system thermodynamically followed by a discussion on the future perspectives of this technology.

3.1 Fundamentals

The first configuration of the CAES system was proposed in the 1940s [1]. This CAES design was, indeed, its most simple configuration based on the definition and objective of this technology, which is storing electricity in the form of high-pressure compressed air energy and then reclaiming it for power generation [2]. Fig. 3.1 presents the schematic of this generation of CAES technology, which was later called diabatic CAES.

As shown in the figure, electricity (coming from any possible source, such as a renewable power plant) is used to drive a compressor with a high–pressure ratio and high capacity to generate compressed air. Then, the compressed air flow, which has been extremely heated up through the compression process, is stored in a reservoir. The reservoir can be an overground tank or chamber with enough strength for the generated pressure or an underground cavern [3]. The latter is much more appropriate for large-scale sizes of a power plant, while the former is applicable for smaller CAES units with lower pressures.

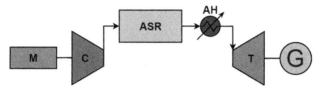

Fig. 3.1 Schematic diagram of a diabatic CAES system; *M*, motor; *C*, compressor; *ASR*, air storage reservoir; *AH*, auxiliary air heater; *T*, turbine; *G*, electricity generator.

The above process is the charging process of a CAES system. Therefore, a diabatic CAES has only three active main components in the charging phase: the motor driving the compressor, the compressor, and the air storage reservoir. On the other hand, the discharging process applies when the stored compressed air is called for electricity generation. For discharging, the diabatic CAES system comprises a high-temperature heater, a single-stage turbine, and an electricity generator. In this stage, the compressed air is heated up to the desired temperature before the air expander. In the past, this heating process was carried out by fossil fuel firing heaters. The air expander is also an individual turbine with a large expansion ratio and capacity. Naturally, expanding the air through the turbine generates rotational power, which is used by the electricity generator for producing power.

The diabatic CAES system can generate a lot of heat in the compression process as a considerable amount of the given electricity to the compressor is converted to heat rather than being successfully converted to work for pressurizing the air stream. However, this heat is simply wasted if the air storage reservoir is not insulated. Therefore, the net efficiency of the energy storage system is not that impressive. The diabatic CAES system may offer an electricity efficiency of about 25%–45% over a roundtrip charging-discharging operation [4].

Considering that the outlet pressurized air of the compressor will be at high temperature and that the compressed air needs to be heated up before expansion, there is the possibility of collecting and storing the thermal energy of the airflow in the charging mode and using it during the discharging process. In this way, the CAES unit would require less auxiliary energy in the heating process and thus it would operate at a greater roundtrip efficiency. For collecting and storing this heating potential, one after-cooler heat exchanger, a pre-heater heat exchanger, and two cold and hot thermal storage units need to be added to the diabatic CAES configuration [5]. The schematic of the new generation of the CAES technology based on this operating strategy, which is called adiabatic CAES, having one or two thermal storage units is shown in Fig. 3.2.

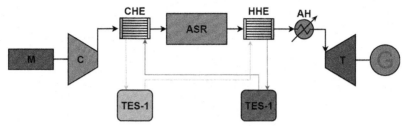

Fig. 3.2 Schematic diagram of an adiabatic CAES system; *CHE*, cooling heat exchanger; *HHE*, heating heat exchanger; *TES*, thermal energy storage.

According to the figure, in an adiabatic single-stage CAES system, in the charging process, the secondary working fluid of the heat exchangers, which is allocated for heat collection of the compression process, is heated up through the cooling heat exchanger and then is poured into the thermal energy storage unit 1 (TES-1). The cooled compressed airflow can then be stored in the air storage reservoir. The secondary working fluid comes from the TES unit, which is at a low temperature. Then, in the discharging mode, this collected heat comes to help reduce the heating duty of the auxiliary air heater by preheating the compressed air flow through the heating heat exchanger. The hot working fluid clearly flows from TES-1 into the heating heat exchanger, and after heating up the airflow and dropping in temperature comes back to the second thermal energy storage unit (TES-2).

Note that the collected heat in an adiabatic CAES system can be stored via different TES methodologies including sensible heat storage (i.e., in a solid material such as a packed bed of rocks or in a thermally stable heat transfer fluid such as industrial oil) and latent heat storage (i.e., molten salt) [6].

From a thermodynamic point of view, compression and expansion turbomachinery, due to the fast process through them as well as the insulated control volumes, present adiabatic processes [7]. Turbomachinery, however, will present its most efficient operation if the expansion and compression actions can be done in isothermal processes. Of course, isothermal expansion and compression would not be possible in practice, but using multistage turbomachinery with intercoolers/interheaters between the stages could push these processes towards semi-isothermal procedures [8]. Fig. 3.3 illustrates how a multistage compression with intercoolers can help for obtaining a better efficiency than a single-stage adiabatic compression process.

As shown, an adiabatic process via a single-stage compressor can approach an isothermal process as the number of compression stages (with

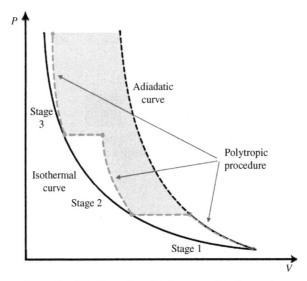

Fig. 3.3 Comparison of adiabatic and multi-stage semi-isothermal compression processes [1].

heat exchangers in between) increases and, in this way, the required work for compressing the same amount of air to the same pressure level decreases significantly. The other advantage of this, as mentioned before, is that the heat gathered in the compression stage can be collected and used for any other applications (e.g., for preheating the airflow before the expansion for a CAES system specifically). Note that the same figure as shown could also be presented for a multi-stage turbine, leading to a greater work production for the same amount of air being expanded in comparison with a single-stage turbine [9].

As such, one could conclude that the next generation of CAES technology might be a multi-stage adiabatic CAES system. This type of CAES system is sometimes simply referred to as an adiabatic CAES, but it might also be called advanced adiabatic CAES or even isothermal CAES [10]. In an isothermal CAES system, the greater the number of stages, the greater the system's efficiency. However, the greater the number of stages, the costlier the system. Therefore, a rigorous techno–economic trade-off depending on the conditions of the specific application of the CAES plant is required to determine the optimal configuration [11]. Fig. 3.4 shows how an isothermal CAES system generally looks.

Another version of the CAES technology, proposed and discussed more recently, is a design in which an isothermal CAES (or advanced adiabatic

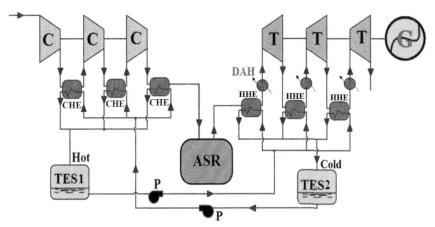

Fig. 3.4 Schematic diagram of an isothermal CAES system coming in a triple-compression/triple-expansion configuration [12].

CAES) comes with no auxiliary fuel-based heater before the expanders. In such systems, the airflow is only heated by the heat collected during the charging process (and possibly an extra low-temperature renewable heat supply system such as a solar thermal unit). In this CAES design, which is called low-temperature CAES, the temperature of airflow before expansion can reach up to 200–250°C [13]. With such a configuration, the roundtrip efficiency of the storage unit would naturally decrease in comparison with that of high-temperature CAES designs. However, the simpler design that is independent from fossil fuels is an advantage [14]. Fig. 3.5 is an illustration of this CAES design with five stages of compression and expansion as well as a solar thermal preheating unit.

An important matter in the operation of CAES systems is the arrangement of the compressors and turbines. The arrangement of the compressors and expanders changes from parallel to series and vice versa as the pressure of the air storage changes through the charging and discharging processes [15]. This is carried out to ensure the best compatibility between the air storage pressure and productivity of the CAES system (producing compressed air in the charging mode and producing power in the discharging mode) to increase efficiency and minimize losses. For the compressors, until the storage pressure is low (lower than the pressure ratio of each of the compressor stages) the stages work in parallel to make a higher mass flow rate. If at this stage the compressors work in series, the outlet pressure of the compressor set will be higher at a lower mass flow rate and this compressed airflow will be expanded in the air storage reservoir, which means wasting the incoming

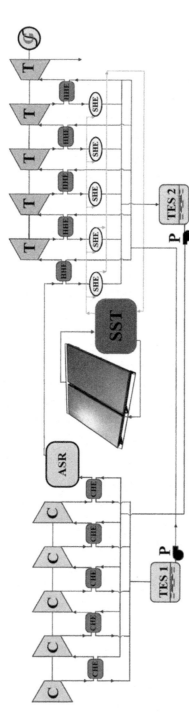

Fig. 3.5 The schematic diagram of a low-temperature CAES system with a solar thermal heater as the excess heat supplier; *SST*, solar storage tank; *SHE*, solar heat exchanger [12].

Table 3.1 Arrangement of compressor and expander stages in different operational pressures of the air storage reservoir.

Air storage pressure (bar)	Compressors arrangement	Turbines arrangement
$P \leq 4$		
$4 < P \leq 16$		
$16 < P \leq 64$		

electricity to drive the compressors. Then, as the pressure of the storage increases, the arrangement changes to a series. This depends on the number of stages, too. For example, if three stages of compression are used, a couple of the stages come in parallel and then become a series with the third one. This can be simply arranged for greater numbers of stages too. In the same manner, for the expanders too, if the air storage reservoir is too high, the expander stages work in series and as the pressure drops, they become parallel. Table 3.1 presents a schematic of the arrangement of the compressors and turbines as the air storage reservoir changes for a triple-stage CAES with a pressure ratio of four for each of the compressors' and expanders' stages [16].

The important point here is that if the air storage reservoir is an underground cavern, the minimum pressure of the reservoir will never drop below a certain level to keep the cavern safe from collapsing under the pressure of the surrounding rocks [17]. In this case, the compressors and expanders will be able to work in series most of the time [18].

3.2 State-of-the-art and practice

In this section, we take a quick glance into the current status of CAES technologies in real-life practice and that of the literature in theory.

3.2.1 State-of-practice

Huntorf power plant is the first pilot-scale CAES unit in the world, which was erected in the late 1970s in Germany [19]. The Huntorf CAES plant is based on a maximum capacity of 60 MW compression process for less than

12 h during the charging phase. The maximum pressure of the air within the cavern is 70 bar. The discharging process is, however, coupled with a gas turbine unit for a nominal capacity of 290 MW for less than 3 h. Indeed, in this power plant, the compressed air stored in the cavern is used as the compressed air required for the combustion process of a gas turbine. Due to the large scale of the plant and the very high maximum pressure of the system, the underground salt cavern has considered for this CAES unit. The Huntorf CAES plant results in a net roundtrip efficiency of 42%. Another important point about the Huntorf CAES plant is its application, which is surprisingly not for stabilizing a renewable power plant's power output, but rather for providing the black-start power of nuclear units and playing the role of an auxiliary service for the local electricity grid. This shows the wide range of possible applications of large-scale agile MES systems, such as CAES technologies. Crotogino [20] presented a very detailed report of the Huntorf CAES plant configuration and characteristics. Table 3.2 summarizes the information in this report.

After the Huntorf CAES unit, the next commercial CAES plant was built in the United States in the early 1990s. The McIntosh CAES plant has a nominal capacity of 110 MW and is able to work at full capacity for 26 h, which is long enough to supply power to more than 110,000 residential buildings [21]. For being able to provide this storage/production capacity, the McIntosh CAES plant has an underground cavern with a volume of 560.000 m^3 with maximum airflow rates of 154 kg/s and 96 kg/s through the

Table 3.2 Main characteristics of the Huntorf CAES plant [20].

Maximum discharging power	290 MW
Maximum discharging duration	3 h
Maximum airflow rate in the expansion process	417 kg/s
Maximum charging power	60 MW
Maximum charging duration	12 h
Maximum airflow rate in the expansion process	108 kg/s
Total air reservoir volume	310,000 m^3
Type of reservoir	Underground cavern
Number of reservoirs	2
Size of reservoirs	140,000 and 170,000 m^3
Minimum operational pressure of reservoirs	20 bar
Maximum operational pressure of reservoirs	70 bar
Highest/lowest depths of the reservoirs into the ground	850/650 m

compression and expansion processes, respectively. In comparison with the Huntorf plant, the McIntosh plant results in a better efficiency due to the waste heat recovery of the exhaust of the combustion process before the expanders. This waste heat recovery unit reduces the fuel consumption of the combustion process by up to 25%, and in this way the net roundtrip efficiency of the McIntosh CAES plant can reach 55%, compared to about 40% in the Huntorf CAES plant [22].

Note that both the McIntosh and Huntorf CAES plants, although being multistage compression and expansion units, are diabatic CAES technologies since the heat generated in their compression units is simply wasted to the atmosphere.

A third pilot-scale CAES plant was supposed to be erected in Iowa Stored Energy Park with a maximum capacity of 270 MW in 2015. This plant was to use aquifer storage instead of caverns, as using water in the aquifer would enable it to regulate the pressure of the air within the reservoir by the constant hydrostatic pressure of the water [23]. This project, however, failed after a long time of feasibility studies and development activities. The main reason for the project failing was the site's geological limitations. This failure, although disappointing, provided practical knowledge about the challenges of such a utility-scale energy storage facility, how to synchronize that with renewable power plants in an independent system marketplace, the economic performance expected from a CAES plant in a real operating condition/market, and so on. [24].

In addition to these plants, there are currently a number of smaller and larger CAES units that are under development/construction around the world. One of these pilot-scale CAES units is the advanced adiabatic CAES unit being constructed in Adele, Germany. The aim of this project is to smooth the power output of local wind farms providing agile backup capacity for a period of 5 h. This project, which has been on hold for a while due to some local market policy conflicts, is supposed to have a maximum storage capacity of 360 MWh with a maximum power output rate of 90 MW. The expected net roundtrip efficiency of this plant is claimed to be about 70% [25]. Some other examples of CAES implantation projects include the Luminant and Shell-Wind Energy CAES project in Texas with 317 MW capacity, Gaelectric Ltd. large-scale CAES plants planned for Northern Ireland and England, and the three CAES projects of the Chinese Academy of Science (with capacities of 1.5 MW, 10 MW, and 100 MW) [26]. Fig. 3.6 presented by Budt et al. [27] gives an illustration of R&D and pilot-scale CAES projects over the history of this technology.

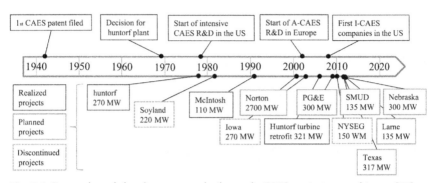

Fig. 3.6 Research and development and pilot-scale CAES projects over history [27].

3.2.2 State-of-the-art

As explained in the previous section, there are not that many pilot-scale CAES plants operating in the world. However, the literature presents thousands of research and review articles focusing on CAES technologies. In this section, we review some of the most recent studies carried out on CAES systems, which result in practical steps forward in the state-of-the-art and practice of this technology.

The studies carried out on CAES are classified into those works optimizing CAES system designs or operation strategies via combination of various CAES designs with different renewable and conventional power plants, and studies proposing/investigating new component technologies including various types of compressors, expanders, heat storage units, or air storage technologies. In respect to the development of CAES literature, one of the most recent review papers by Budt et al. [27] examines in detail the basic principles of CAES systems, the past milestones of the technology, and the most recent developments in this framework.

Most of the studies on CAES systems are investigations or feasibility studies on the use of CAES units together with renewable or conventional power plants, that is, the concept of integrated CAES systems. An integrated CAES is, indeed, a combined system of CAES-renewable power plant where the surplus electricity produced from the renewable power plant (e.g., a wind turbine) is used to generate compressed air to be stored and then used for electricity generation [28]. For example, a combination of a CAES unit with a PV plant in the north of Brazil was investigated techno-economically by Arabkoohsar et al. [2]. The same team of researchers also carried out a feasibility study on combining the hybrid system with the local power productive gas station [14]. A combination of a CAES unit with a wind farm for the sake of wind power fluctuation mitigation was analyzed by Pei and Zheng [29], resulting in making opportunities for supplying more wind power with more stability to the local grid. A combined CAES-solar power-gas-based combined cooling, heat and power (CCHP) plant was designed and optimized by Wang et al. [30]. The off-design performance analysis of this system was carried out later by the same team to see how partial load operation may affect the productivity and efficiency of the hybrid system [31]. Studies proposing the combination of CAES systems with other energy systems are also found in the literature. A combined CAES-multi-effect seawater desalination (MED) system is an example of such works [32].

A review of expanders to be used in different CAES designs is presented in [33]. The conclusion of this study is that more efficient and new designs of air expanders are required for high-pressure, high-temperature CAES systems in order to minimize pressure losses and increase energy efficiency. The study also found that improving the capacity of volumetric expanders (such as reciprocating expanders, single and twin-screw expanders, scroll expanders, etc.) improves the cost-effectiveness of small-scale CAES systems. The article presents a very informative table that indicates which type of expander (including reciprocating machines, screw or scroll expanders as rotatory positive displacement machines and radial and axial expanders as turbomachines) for which CAES technology and capacity is most appropriate [33]. Lemort et al. [34] presented a detailed comparison of piston, screw, and scroll expanders for power cycles and found that reciprocating turbines are more appropriate for micro- and small-scale CAES systems due to high internal volume ratios that range from six to fourteen. This is because in such scaled-down CAES units, higher pressures are required to enhance power density and energy production due to the small storage capacity and low flow rate. Saadat et al. [35] introduced an innovative liquid-piston expander suitable for CAES systems that enhances the energy and power density of air storage via a near isothermal expansion process. Rotatory expanders including both screw and scroll expanders have also been studied for CAES systems. Single and double-screw expanders were investigated for power production systems including CAES systems in [36, 37]. Compared to reciprocating expanders, screw expanders usually offer lower pressure ratios because of their smaller built-in volume ratios, and their isentropic efficiency can range from 20% to 70% [38]. Such air turbines are appropriate for CAES systems as they allow two-phase working fluid, though they are not recommended for capacities less than 10 kW

[34]. Alsagri et al. [39] proposed the use of such compressors in lower temperature CAES technologies due to their greater operation stability at low temperatures compared to that of regular compressors. Al Jubori et al. [40] analyzed the effects of using efficient radial-inflow expanders on the performance of small-scale CAES systems using mean-line design and computational fluid dynamic (CFD) techniques and observed a maximum expander efficiency of 76.70%.

Centrifugal air compressors were proposed to be used in CAES systems with large-scale applications in [1, 6, 15]. Performance analysis of such compressors (i.e., linear reciprocating compressors) in CAES systems is presented in Ref. [41]. According to Ref. [42], for compressing the inlet air of a large-scale CAES unit (like the Huntorf and McIntosh plants currently in operation) either axial or radial compressors might be used. In the case of an axial compressor, a lower pressure ratio and a higher flow rate would be achieved, while with a radial compressor a lower flow rate and higher compression ratio might be obtained. Wieberdink et al. [43] investigated the use of high-pressure (up to 210 bar) liquid-piston air compressors (and expanders) in adiabatic CAES systems as well as the impacts of porous media insert on the system's efficiency and power density. Iglesias and Favrat [44] analyzed the use of isothermal oil-free co-rotating scroll compressors (and expanders) in CAES systems and brought an update of the thermodynamic model of scroll compressors taking water injection into account. The use of scroll compressors for micro-CAES systems was investigated by Ma et al. [45]. They also carried out an extensive energy efficiency assessment to estimate the optimal operation range of the compressor in the CAES unit. Heidari et al. [46] designed, modeled, and experimentally tested an innovative finned reciprocating (piston) compressor for isothermal CAES systems observing a 42% improvement of the efficiency of the system in comparison with the classical piston.

A review of studies on the thermal storage methods/technologies to be employed in CAES systems is presented in [47]. The possible TES methods are sensible, latent, and thermochemical heat storage techniques, among which sensible heat storage is the most straightforward, simple, and mature method [48]. A tank of working fluid (full of pressurized water or heat transfer oil) [49] that increases in temperature as it receives heat during the charging mode is the conventional manner of heat storage in CAES systems. Another possible solution in the category of sensible heat storage for CAES systems is a packed bed of rocks [50]. Barbour et al. [51] analyzed the performance of a CAES system accompanied by a packed bed of stones as the thermal storage unit. Want et al. [52] analyzed the performance of a CAES plant accompanied by a packed bed of stones and an electrical heater for co-storage of power and heat. On the other hand, latent thermal energy storage and thermochemical heat storage techniques have generally received much attention over the last years due to their very higher storage density compared to that of sensible heat storage methods [53].

As mentioned, both underground and aboveground air storage systems are possible options for CAES systems. The former is more appropriate for large-scale CAES systems, while the latter is suitable for small-scale units. The underground storage is usually a salt or rock cavern very deep in the ground, although other forms of geological structures have also come into consideration. For example, a concrete-lined tunnel has been recently tested as the air storage reservoir of a 2-MW CAES system in Japan [54]. Other examples of such efforts include the test facility made by the Electric Power Research Institute (EPRI) using hard rock caverns with water compensation [55], and the 25 MW porous rock-based CAES unit made by Enel in Italy [56]. The use of natural gas pipelines for compressed air storage (as an alternative for underground storage) was also proposed and investigated by Valdivia et al. [57] in Chile. Li et al. [58] proposed and analyzed the use of artificial, low-permeability barriers in high-permeability aquifers to improve CAES system performance. They concluded that grout viscosity can have impacts on the distribution of permeability, that the special gravity of grout might have severe effects on the shape of the barrier, and that the influence of water injection on the behavior of the barrier is insignificant. Pimm et al. [59] proposed and designed a system called energy bags, which are fabric vessels located at a suitable depth of the sea for underwater compressed air storage. Indeed, an underwater CAES system takes advantage of the hydrostatic pressure and unlimited space under the sea. The volume of the air storage bag in this system changes during the charge and discharge processes so that nearly constant pressure is always achieved. This is beneficial for keeping the compressors and turbines at their nominal design pressure operating conditions. The disadvantage of this method of air storage is the limited maximum pressure that can be obtained due to the limited height of the seawater.

In addition to the aforementioned CAES designs, there are a number of CAES designs that have been proposed, mainly via improving the configuration of existing systems. In this regard, LightSail Energy Ltd. [60], a company working on CAES technologies, has modified the design of the adiabatic CAES system by employing reversible reciprocating piston turbomachinery, which results in a nearly isothermal process. Arabkoohsar et al. [16] proposed the novel concept of the subcooled CAES system, which is appropriate for the trigeneration of heat, cold, and power at high overall efficiency. They investigated the use of this CAES system for various applications [15, 18] along with the system's partial load operation performance [49]. As the power-to-power efficiency of subcooled CAES design is less than other CAES designs, the combination of a subcooled CAES system with a small-scale ORC unit to increase the electrical efficiency of the system was investigated in [39]. The subcooled CAES system was further investigated in several other studies in which it was referred to as trigeneration CAES [61–64].

Besides these, there are a number of older and newer designs of energy storage that are partly similar to CAES technologies. Although these do not work based on the same exact principles as CAES systems, we discuss them here because of the similarity of their concepts to CAES. One of these technologies is compressed-air storage with humidification (CASH). A CASH system is, in fact, a CAES system with an air saturator (to humidify the airflow before the expansion) and in this way results in improved roundtrip efficiency. A detailed explanation of this technology is presented in [65]. Liquid air energy storage (LAES) system is another version of CAES that stores liquid air (or liquid nitrogen) instead of compressed air, providing large-scale energy storage possibility at a smaller size than a CAES system [66]. A step forward in the technology of LAES with the aim of increasing the overall efficiency is supercritical CAES (SC-CAES) system [67]. The SC-CAES system offers greater efficiency than the LAES system (about 70%) without being restricted to existing air liquefaction technology limitations and an extremely larger energy density than the regular CAES technologies (i.e., almost 18 times larger) [67].

3.3 Mathematical model

In this section, we present the required mathematical formulations for general modeling of a CAES unit. This includes the required model for energy analysis of the system. Naturally, a variety of different analyses and investigations (e.g., energy analysis, exergy and exergoeconomic assessments, economic analysis, optimizations of design and operation strategy, geological investigations, etc.) are required to be carried out on a CAES system to determine its feasibility, optimal design, and operation management; however, these are all out of the scope of this book, which is focused on the technical and general aspects of CAES technologies.

3.3.1 General principles

For an energy analysis of the system, one needs to know the first law of thermodynamics (also called the energy conservation law). The general format of this law for a control volume is as [68]:

$$\frac{dE_{C.V}}{dt} = \dot{Q}_{c.v} + \dot{W}_{c.v} + \sum \dot{m}_i \left(h_i + \frac{Ve_i^2}{2} + gz_i \right)$$
$$- \sum \dot{m}_e \left(h_e + \frac{Ve_e^2}{2} + gz_e \right) \quad (3.1)$$

In which $\dot{Q}$ is the rate of heat transfer from/into the control volume, $\dot{W}$ is the rate of work carried out by/on the control volume, $\dot{m}$ is the mass flow rate of the working fluid through the control volume, h is the specific enthalpy of the working fluid, Ve is the velocity of the working fluid, and z is the potential term of the control volume. The parameter $E_{C.V}$ represents the control volume's total energy, which comprises the internal, kinetic, and potential energy terms. Also, the subscript "I" represents the internal condition of the control volume and the subscript "e" represents the outlet state of the control volume.

Besides the energy conservation law, most of the time the mass conservation law is also required to be able to fulfill the energy analysis of a system. The general format of this law for a control volume is:

$$m^{\lambda+1} = \left(\sum \dot{m}_i - \sum \dot{m}_e \right)^{\lambda} \Delta t + m^{\lambda} \quad (3.2)$$

Where m and λ represent the instantaneous mass of the control volume and the operational time-steps of the control volume, respectively.

Finally, the energy efficiency of a system, also known as the first law efficiency, can be defined as:

$$\xi = \frac{\text{Useful energy output}}{\text{Energy input}} \quad (3.3)$$

This correlation comes from the general understanding of the efficiency of a given asset, which is defined as the portion of the input parameter that is successfully converted/gained/produced as the output of the asset.

The energy analysis model of the most efficient CAES design, that is, isothermal CAES or advanced adiabatic CAES system, is presented hereunder. For this, first, the allowed assumptions that can be considered for simplifying the energy modeling of the isothermal CAES in a reasonable manner are discussed.

The first fact is that the airflow through the CAES system operation, most of the time, can be considered as an ideal gas. This is, naturally, a matter of the maximum pressure and lowest temperature that the airflow might go when the CAES system is operating. The higher the pressure and the lower the temperature of the gas, the more it deviates from an ideal gas state. In general, for considering a gas (or gas mixture) flow as an ideal gas, the compressibility factor of the gas should be close to uniform. The compressibility factor of a gas can be estimated by the following correlation:

$$Z = 1 + \frac{B}{v'_r} + \frac{C}{v'^2_r} + + \frac{D}{v'^5_r} + \frac{c_4}{T^3_r v'^2_r} + \left(\lambda + \frac{\xi}{v'^2_r}\right) \exp\left(-\frac{\xi}{v'^2_r}\right) \quad (3.4)$$

where the terms B, C, D, v_r, and T_r are given as follows:

$$B = b_1 - \frac{b_2}{T_t} - \frac{b_3}{T^2_t} - \frac{b_4}{T3}; \; C = c_1 - \frac{c_2}{T_t} + \frac{c_3}{T^3_t}; \; D = d_1 \frac{d_2}{T_t}; \; T_t = \frac{T}{T_c};$$
$$v'_t = \frac{v}{RT_c / P_c} \quad (3.5)$$

In the preceding correlation set, the parameters v, T_c, and P_c, represent the specific volume of the gas, the critical temperature of the gas, and the critical pressure of the gas, respectively. The constant coefficients $b1:b4$, $c_1:c_4$, $d_1:d_2$, λ, and ζ are given by the following Table 3.3.

The investigations show that in the common ambient temperature level and up to 200 bar, the airflow can be considered as an ideal gas without bringing any major error into the calculations.

The other technical assumptions possible to make in this context are [68]:

Table 3.3 The constant coefficients of the compressibility factor correlation.

Constant	Value	Constant	Value
b_1	0.1181193	c_3	0
b_2	0.265728	c_4	0.042724
b_3	0.154790	d_1	0.0000155488
b_4	0.030323	d_2	0.0000623689
c_1	0.0236744	λ	0.65392
c_2	0.0186984	ζ	0.060167

- Air humidity effects are negligible because ambient air is not too humid in many conditions, otherwise the CAES systems will be furnished with adequate dehumidifiers.
- CAES systems are agile energy storage technologies, meaning that their start-up time is quite short. If the modeling is to be dynamic, the start-up time should be calculated based on the specific characteristics of the turbomachinery of the plant and the other system components. Otherwise, a fast approach to the steady-state operating condition can be reasonable.
- The expanders, heat exchangers, and storage tanks in the CAES system are assumed to be perfectly insulated.
- The temperature of compressed air in the cavern is assumed to be equal to the mean temperature of the cavern rocks, which is a function of the depth of the cavern underground. This is also an acceptable assumption because not only will intercoolers and the aftercooler collect almost all of the heat generated in the airflow before it is stored in the cavern, but also the compressed airflow will have enough time to reside in the cavern and approach a uniform temperature with its environment.
- The variations of kinetic and potential energies through the turbomachinery and the heat exchangers are negligible.

Note that mechanical systems are designed for certain operating conditions and capacity. Therefore, as the operational state or the load of operation deviates from the nominal conditions, the performance of the components and, consequently, the whole system is negatively affected. Therefore, it is not reasonable to neglect the effect of off-design operation conditions on the performance of the system unless the system either in charging or discharging mode does not fall below a certain minimum-allowed operational load level.

3.3.2 Thermodynamic model

For an isothermal CAES, like the one shown in Fig. 3.4, the thermodynamic model can come in two different phases of charging and discharging, each of which has a number of control volumes.

In the charging mode of a CAES system, the compression unit, the intercoolers, the aftercooler, the thermal storage units, and the cavern are the active parts of the system.

For the compressors, the net mechanical work of the turbomachinery used for compressed air production will be equal to the supplied electricity flow (P_E) multiplied by the compressors' mechanical efficiency (η_C), which is so close to unity [69].

$$\dot{W}_C = P_E \times \eta_C \tag{3.6}$$

Knowing the compressors' isentropic efficiency and based on simple thermodynamic formulations, the specific actual work of each of the compressor stages (w_{c-act}) could be calculated [68]. Then, the compressor stage outlet air temperature might be calculated by:

$$T_{c-e} = \frac{w_{c-act} + c_{p-i}T_{c-i}}{c_{p-e}} \tag{3.7}$$

In which, T_{c-i}, c_{p-i}, and c_{p-e} are the compressor stage inlet air temperature and the specific heat capacity of air at the inlet and outlet conditions of the compressor, respectively.

Having the specific work of the compressors, the following equation could be used for calculating the total and individual mass flow rate of compressors:

$$\dot{W}_C = \sum_{j=1}^{N} (\dot{m}_c w_c)_j; \; \dot{m}_C = \sum_{j=1}^{N} \dot{m}_{cj} \tag{3.8}$$

Here, c refers to each of the compressor stages, C represents the whole compressor set, and N is the number of stages of the compressor.

To calculate the specific work of the compressor stages, one needs to know the inlet temperature of each stage. For this, the intercoolers should be modeled simultaneously with the compressors. For the intercoolers and the aftercooler, one can write [70]:

$$q_{che} = U_{che} - A_{che} - \Delta T_{lm} \tag{3.9}$$

q_{che} is the rate of heat being transferred from the airflow to the secondary fluid through the heat exchanger. A_{che}, U_{che}, and ΔT_{lm} are the heat transfer area, the overall heat transfer coefficient, and the logarithmic temperature difference, respectively. The latter two could be calculated as follows:

$$U_{che} = \frac{1}{{}^{1}/_{h_a} + {}^{1}/_{h_f}} \tag{3.10}$$

$$\Delta T_{lm} = \frac{(T_{i-a} - T_{i-f}) - (T_{e-a} - T_{e-f})}{Ln\left(\dfrac{(T_{i-a} - T_{i-f})}{(T_{e-a} - T_{e-f})}\right)} \tag{3.11}$$

Here, h_a and h_f are the convective heat transfer coefficients of the airflow and the secondary working fluid, $T_{i\text{-}a}$ and $T_{i\text{-}f}$ are the inlet temperatures of the air and secondary working fluid, and $T_{e\text{-}a}$ and $T_{e\text{-}f}$ are the outlet temperatures of the airflow and the working fluid from the heat exchanger.

Calculating the convective heat transfer coefficients is a matter of the regime of the flows. The airflow is certainly a turbulent flow, but the working fluid could be laminar if the operational load is too low (which is of course very unlikely if the system is not going to fall into too-low operational loads). For the laminar and turbulent flows, the following two correlations could be used to calculate the convective heat transfer coefficient, respectively [71]:

$$h = 0.023 \times \mathrm{Re}^{0.8} \times \mathrm{Pr}^{0.4} \times k/D_h \tag{3.12}$$

$$h = {}^{(4.63 \times k)}/_{D_h} \tag{3.13}$$

In which Re, Pr, k, and D_h are the Reynolds number, the Prandtl number, the thermal conductivity factor of the fluid, and the hydraulic diameter of the channel of the flow, respectively.

Finally, using the preceding correlations, one could use the following formulas to calculate, respectively, the outlet temperatures of the airflow and the secondary working fluid from the intercoolers and the aftercooler:

$$T_{e-a} = T_{i-a}(1 - \varepsilon) + \varepsilon \cdot T_{i-f} \tag{3.14}$$

$$T_{e-f} = \frac{\varepsilon \cdot \dot{m}_a \cdot c_{p-a} \cdot \left(T_{i-a} - T_{i-f}\right)}{\dot{m}_f \cdot c_f} + T_{i-f} \tag{3.15}$$

Here, ε is defined as the effectiveness factor of the heat exchangers and can be calculated by [72]:

$$\varepsilon = \frac{UA/C_{\min}}{1 + UA/C_{\min}}; C_{\min} = \dot{m}_a c_{p-a} \tag{3.16}$$

In which UA is the overall heat transfer coefficient (considering the area of heat transfer of the heat exchanger) and $C_{\min}$ is the lower specific heat value among the two fluids through the heat exchanger (i.e., the airflow and the secondary working fluid), which evidently belongs to the airflow.

Naturally, for calculating the preceding set of formulas, as the correlations are dependent on each other, an iterative solution is required. Implementing this iterative solution, the mass flow rate of the generated airflow, the temperature of that at each stage, and the temperature and mass flow rate of the secondary working fluid could be calculated.

Then, for the heat storage tanks, which are both active in the charging phase, based on the first law of thermodynamics (assuming a lumped control volume) and the mass conservation law, one can write [73]:

$$T_{st}^{\lambda+1} = \frac{\left(\dot{m}_i c_f T_i - \dot{m}_e c_f T_e\right)^{\lambda} \Delta t + m_{st}^{\lambda} c_f T_{st}^{\lambda}}{m_{st}^{\lambda+1} c_f} \tag{3.17}$$

$$m_{st}^{\lambda+1} = m_{st}^{\lambda} + \left(\dot{m}_i - \dot{m}_e\right)_f^{\lambda} \Delta t \tag{3.18}$$

Here m_{st} is the instantaneous total mass of the thermal storage units.

Finally, for the air storage cavern, based on the mass conservation law, one has:

$$m_{ca}^{\lambda+1} = m_{ca}^{\lambda} + \left(\dot{m}_i - \dot{m}_e\right)_a^{\lambda} \Delta t \tag{3.19}$$

Having V_{ca} as the volume of the cavern and T_{ca} as its mean temperature, the instantaneous pressure of the cavern is calculated by:

$$P_{ca} = {}^{m_{ca} R T_{ca}} \big/ {}_{V_{ca}} \tag{3.20}$$

Discharging phase: The facilities activating in the discharging phase of a CAES unit are the preheater and interheater heat exchangers, the auxiliary heaters, the air expanders, the cavern, and the thermal storage units.

For the air expanders, the amount of work required to be produced at any instant is given by [74]:

$$\dot{W}_T = {}^{P_G} \big/ {}_{\eta_{TG}} \tag{3.21}$$

In which P_G is the rate of power to be produced by the CAES system and η_{TG} is the total efficiency of the turbo-generator set converting the mechanical power of the turbines to electricity.

Similar to the compressor stages, for calculating the actual outlet temperature of the turbines, the actual work of the turbine should be calculated using information about the inlet conditions, the isentropic efficiency of the turbines, and the isentropic outlet conditions. Then, one has:

$$T_{t-e} = \frac{w_{t-\text{act}} + c_{p-i} T_{t-i}}{c_{p-e}} \tag{3.22}$$

Then, using the same correlations as those used for the intercoolers for the preheater and the interheaters, and taking an iterative solution, one would be able to calculate the work production of the turbines and the required mass flow rate through each of the turbine stages and, consequently, the total airflow rate and turbine work. For this, one has:

$$\dot{W}_T = \sum_{j=1}^{N} (\dot{m}_t w_t)_j; \ \dot{m}_T = \sum_{j=1}^{N} \dot{m}_{tj} \qquad (3.23)$$

Understanding the situation of the airflow after the interheaters and the design inlet conditions for the turbines, one could calculate the heating duty of the auxiliary haters and the amount of fuel required for them to be burnt. The following two equations, respectively, calculate the heating duty of each of the heaters $(\dot{Q}_h)$ and the required fuel for each of them $(\dot{V}_{fu})$ at any instance [75].

$$\dot{Q}_h = \dot{m}_a c_{p-a} (T_{t-i} - T_{h-i}) \qquad (3.24)$$

$$\dot{V}_{fu} = \frac{\dot{Q}_h}{LHV \, \eta_h} \qquad (3.25)$$

In which T_{h-i} is the inlet temperature of the air into the heater and T_{t-i} is the required temperature before the turbine, which is a constant temperature based on the design setpoint of the turbine. Also, LHV is the fuel lower heating value and η_h is the heater efficiency.

It is again mentioned that the same correlations as those used in the charging phase modeling can be used here for the air storage cavern and the thermal storage units, too.

Having the preceding equations, one could simply calculate the net first law efficiency of the CAES unit over a roundtrip operation as:

$$\eta = \frac{\sum_{t=1}^{tdc} P_G}{\sum_{t=1}^{tc} P_E + \sum_{t=1}^{tdc} \dot{Q}_h} \qquad (3.26)$$

Where tc shows the number of time-steps the system is on the charging mode and tdc is the number of discharging time-steps until the cavern is fully discharged to the initial condition at the beginning of the charging mode.

3.4 Future perspective

According to the literature and state-of-practice presented for the CAES technologies in the previous sections, one can clearly see that there is a very big gap between the point that the research and development studies have reached and what is available in the market and the real-life application of this MES technology. For this, a number of reasons could be addressed.

- Part of this can be due to the fact that energy storage technologies have not really been critically demanded so far as renewable energy technologies are not dominating the global energy systems. In fact, energy storage technologies are vitally needed when more and more fluctuating power is supplied to the grids and, consequently, stabilizing the grid is challenging.
- The other reason for this is, surely, the well-developed state-of-practice of batteries, which are commonly used in renewable power plants (although being costly), while CAES technology is not still perfectly developed. Indeed, most of the advancements in CAES systems and the progress in the more efficient designs of CAES technology have occurred only very recently.
- The certain geographical conditions required for constructing salt/rock caverns (which is necessary for large-scale CAES units) is really a restriction. This problem, for example, stopped the Iowa large-scale CAES plant project.

Today, however, renewable energy plants, specifically wind and solar power plants, are getting more interest worldwide and the wind/solar power share of electricity grids of countries is instantly increasing. This means that the penetration of fluctuating power to the grids is increasing and therefore there must be some efficient and reliable storage facilities to control the frequency and increase the reliability of the electricity grids.

The required investment of CAES technology is not too high, and compared to batteries it requires much less investment for the same storage capacity. With such a fairly low capital investment required, and considering the operation and maintenance costs and the useful lifetime of a CAES system (which can be at least 20–25 years), the levelized cost of CAES systems, especially advanced designs, is much less than many other energy storage technologies. CAES is among the very few electricity storage technologies that can be sized for storage time frames ranging from a couple of hours to a number of days or even weeks and from low capacities to very large scales.

This is a big advantage that CAES systems offer. In addition, the agility of the technology (which is extremely important in real-time frequency stabilization of the grid) and the acceptable roundtrip efficiency of the system are further advantages of CAES technologies.

Taking into account the aforementioned key reasons for the slow deployment of CAES technology in practice and the advantages that a CAES system offers, as well as the fast motion of the world's energy matrix towards a 100% renewable energy supply, there is much hope that CAES units will come into service in the near future. As mentioned, there are also new promising designs of CAES systems that offer co- or multi-generation and thereby can pave the road for the integration of different energy sectors. This feature of new and emerging CAES designs can also be very beneficial for future smart energy systems. The expectation is that the literature will reach a more advanced so that these highly efficient and cost-effective systems can be demonstrated on a pilot scale and finally be successfully launched to the market.

References

[1] A. Arabkoohsar, Dynamic Modeling of a Compressed Air Energy Storage System in a Grid Connected Photovoltaic Plant, PhD Thesis(2016).

[2] A. Arabkoohsar, L. Machado, M. Farzaneh-Gord, R.N.N. Koury, Thermo-economic analysis and sizing of a PV plant equipped with a compressed air energy storage system. Renew. Energy 83 (2015), https://doi.org/10.1016/j.renene.2015.05.005.

[3] T. Wang, C. Yang, H. Wang, S. Ding, J.J.K. Daemen, Debrining prediction of a salt cavern used for compressed air energy storage. Energy 147 (2018) 464–476, https://doi.org/10.1016/j.energy.2018.01.071.

[4] B. Elmegaard, W. Brix, Efficiency of compressed air energy storage, in: Proceedings of 24th International Conference of Efficiency, Cost, Optimization, Simulation and Environmental Impact of Energy Systems ECOS 2011, 2011, pp. 2512–2523.

[5] H. Peng, Y. Yang, R. Li, X. Ling, Thermodynamic analysis of an improved adiabatic compressed air energy storage system. Appl. Energy 183 (2016) 1361–1373, https://doi.org/10.1016/j.apenergy.2016.09.102.

[6] A. Arabkoohsar, G.B.B. Andresen, Design and analysis of the novel concept of high temperature heat and power storage. Energy 126 (2017) 21–33, https://doi.org/10.1016/j.energy.2017.03.001.

[7] J.P. Stark, Fundamentals of classical thermodynamics (Van Wylen, Gordon J.; Sonntag, Richard E.). J. Chem. Educ. 43 (1966) A472, https://doi.org/10.1021/ed043p A472.1.

[8] A. Arabkoohsar, L. Machado, M. Farzaneh-Gord, R.N.N. Koury, The first and second law analysis of a grid connected photovoltaic plant equipped with a compressed air energy storage unit. Energy 87 (2015) 520–539, https://doi.org/10.1016/j.energy.2015.05.008.

[9] C.A. Hall, S.L. Dixon, Fluid Mechanics and Thermodynamics of Turbomachinery, sixth ed., (2010).

[10] F. Jabari, S. Nojavan, B. Mohammadi Ivatloo, Designing and optimizing a novel advanced adiabatic compressed air energy storage and air source heat pump based

μ-combined cooling, heating and power system. Energy 116 (2016) 64–77, https://doi.org/10.1016/j.energy.2016.09.106.

[11] A. Arabkoohsar, G.B. Andresen, Dynamic energy, exergy and market modeling of a high temperature heat and power storage system. Energy 126 (2017), https://doi.org/10.1016/j.energy.2017.03.065.

[12] A. Arabkoohsar, L. Machado, R.N.N. Koury, K.A.R. Ismail, Energy consumption minimization in an innovative hybrid power production station by employing PV and evacuated tube collector solar thermal systems. Renew. Energy 93 (2016) 424–441, https://doi.org/10.1016/j.renene.2016.03.003.

[13] D. Wolf, M. Budt, LTA-CAES—a low-temperature approach to adiabatic compressed air Energy storage. Appl. Energy 125 (2014) 158–164, https://doi.org/10.1016/j.apenergy.2014.03.013.

[14] A. Arabkoohsar, L. Machado, R.N.N. Koury, Operation analysis of a photovoltaic plant integrated with a compressed air energy storage system and a city gate station. Energy 98 (2016) 78–91, https://doi.org/10.1016/j.energy.2016.01.023.

[15] A. Arabkoohsar, An integrated subcooled-CAES and absorption chiller system for cogeneration of cold and power, in: IEEE Xplore, Proceeding SEST, 2018 2018, pp. 1–5.

[16] A. Arabkoohsar, M. Dremark-Larsen, R. Lorentzen, G.B. Andresen, Subcooled compressed air energy storage system for coproduction of heat, cooling and electricity. Appl. Energy 205 (2017) 602–614, https://doi.org/10.1016/j.apenergy.2017.08.006.

[17] M. Raju, S. Kumar Khaitan, Modeling and simulation of compressed air storage in caverns: a case study of the Huntorf plant. Appl. Energy 89 (2012) 474–481, https://doi.org/10.1016/j.apenergy.2011.08.019.

[18] A. Arabkoohsar, G.B. Andresen, Design and optimization of a novel system for trigeneration. Energy 168 (2019) 247–260, https://doi.org/10.1016/j.energy.2018.11.086.

[19] S. Briola, P. Di Marco, R. Gabbrielli, J. Riccardi, A novel mathematical model for the performance assessment of diabatic compressed air energy storage systems including the turbomachinery characteristic curves. Appl. Energy 178 (2016) 758–772, https://doi.org/10.1016/j.apenergy.2016.06.091.

[20] F. Crotogino, K.-U. Mohmeyer, R. Scharf, Huntorf CAES: more than 20 years of successful operation, in: Solution Mining Research Institute Spring Meeting, 2001, pp. 351–357.

[21] McIntosh CAES Plant. http://www.powersouth.com/wp-content/uploads/2017/07/CAES-Brochure-FINAL.pdf.

[22] I. Arsie, V. Marano, M. Moran, G. Rizzo, G. Savino, Optimal Management of a Wind/CAES Power Plant by Means of Neural Network Wind Speed Forecast, (2007).

[23] C. Guo, L. Pan, K. Zhang, C.M. Oldenburg, C. Li, Y. Li, Comparison of compressed air energy storage process in aquifers and caverns based on the Huntorf CAES plant. Appl. Energy 181 (2016) 342–356, https://doi.org/10.1016/j.apenergy.2016.08.105.

[24] Microsoft Word—Lessons Learned (SAND2012-0388) v4_Enhanced Reader.pdf.

[25] RWE Power, Rheinisch Westfälisches Elektrizitätswerk AG, ADELE–Adiabatic Compressed-Air Energy Storage for Electricity Supply, Köln, (2010) pp. 4–5. https://www.rwe.com/web/cms/mediablob/en/391748/data/364260/1/rwe-power-ag/innovations/Brochure-ADELE.pdf.

[26] https://estoolbox.org/index.php/en/background-2/8-samples/9-caes-introduction.

[27] M. Budt, D. Wolf, R. Span, J. Yan, A review on compressed air energy storage: basic principles, past milestones and recent developments. Appl. Energy 170 (2016) 250–268, https://doi.org/10.1016/j.apenergy.2016.02.108.

[28] S.D. Garvey, The dynamics of integrated compressed air renewable energy systems. Renew. Energy 39 (2012) 271–292, https://doi.org/10.1016/j.renene.2011.08.019.

[29] H. Jin, P. Liu, Z. Li, Dynamic modeling and design of a hybrid compressed air energy storage and wind turbine system for wind power fluctuation reduction. Comput. Chem. Eng. 122 (2019) 59–65, https://doi.org/10.1016/j.compchemeng. 2018.05.023.

[30] X. Wang, C. Yang, M. Huang, X. Ma, Multi-objective optimization of a gas turbine-based CCHP combined with solar and compressed air energy storage system. Energy Convers. Manag. 164 (2018) 93–101, https://doi.org/10.1016/j.enconman. 2018.02.081.

[31] X. Wang, C. Yang, M. Huang, X. Ma, Off-design performances of gas turbine-based CCHP combined with solar and compressed air energy storage with organic Rankine cycle. Energy Convers. Manag. 156 (2018) 626–638, https://doi.org/10.1016/j. enconman.2017.11.082.

[32] A. Razmi, M. Soltani, M. Tayefeh, M. Torabi, M.B. Dusseault, Thermodynamic analysis of compressed air energy storage (CAES) hybridized with a multi-effect desalination (MED) system. Energy Convers. Manag. 199 (2019) 112047, https://doi.org/10.1016/ j.enconman.2019.112047.

[33] W. He, J. Wang, Optimal selection of air expansion machine in compressed air energy storage: a review. Renew. Sust. Energ. Rev. 87 (2018) 77–95, https://doi.org/ 10.1016/j.rser.2018.01.013.

[34] V. Lemort, L. Guillaume, A. Legros, S. Declaye, S. Quoilin, A Comparison of Piston, Screw and Scroll Expanders for Small Scale Rankine Cycle Systems, (2013).

[35] M. Saadat, P.Y. Li, Combined optimal design and control of a near isothermal liquid piston air compressor/expander for a compressed air energy storage (CAES) system for wind turbines. in: ASME 2015 Dynamic System and Control Conference DSCC 2015, American Society of Mechanical Engineers, 2015, https://doi.org/10.1115/ DSCC2015-9957.

[36] W. He, Y. Wu, Y. Peng, Y. Zhang, C. Ma, G. Ma, Influence of intake pressure on the performance of single screw expander working with compressed air. Appl. Therm. Eng. 51 (2013) 662–669, https://doi.org/10.1016/j.applthermaleng.2012.10.013.

[37] M. Read, N. Stosic, I.K. Smith, Optimization of screw expanders for power recovery from low-grade heat sources. Energy Technol. Policy 1 (2014) 131–142, https://doi. org/10.1080/23317000.2014.969454.

[38] W. Wang, Y.T. Wu, G.D. Xia, C.F. Ma, J.F. Wang, Y. Zhang, Experimental study on the performance of the single screw expander prototype by optimizing configuration. in: ASME 2012 6th International Conference of Energy Sustainability ES 2012, collocated with ASME 2012 10th International Conference of Fuel Cell Science Engineering and Technology, American Society of Mechanical Engineers, 2012, pp. 1281–1286, https://doi.org/10.1115/ES2012-91502.

[39] A.A. Alrobaian, A.S. Alsagri, A. Arabkoohsar, Combination of subcooled compressed air energy storage system with an organic rankine cycle for better electricity efficiency, a thermodynamic analysis, J. Clean. Prod. 239 (2019) 118119.

[40] A.M. Al Jubori, Q.A. Jawad, Investigation on performance improvement of small scale compressed-air energy storage system based on efficient radial-inflow expander configuration. Energy Convers. Manag. 182 (2019) 224–239, https://doi.org/10.1016/j. enconman.2018.12.048.

[41] M. Heidari, S. Wasterlain, P. Barrade, F. Gallaire, A. Rufer, Energetic macroscopic representation of a linear reciprocating compressor model. Int. J. Refrig. 52 (2015) 83–92, https://doi.org/10.1016/j.ijrefrig.2014.12.019.

[42] D.A. Katsaprakakis, Energy storage for offshore wind farms. in: Offshore Wind Farms: Technologies Design and Operation, Elsevier Inc, 2016, pp. 459–493, https://doi.org/ 10.1016/B978-0-08-100779-2.00015-5.

[43] J. Wieberdink, P.Y. Li, T.W. Simon, J.D. Van de Ven, Effects of porous media insert on the efficiency and power density of a high pressure (210 bar) liquid piston air

compressor/expander—an experimental study. Appl. Energy 212 (2018) 1025–1037, https://doi.org/10.1016/j.apenergy.2017.12.093.

[44] A. Iglesias, D. Favrat, Innovative isothermal oil-free co-rotating scroll compressor-expander for energy storage with first expander tests. Energy Convers. Manag. 85 (2014) 565–572, https://doi.org/10.1016/j.enconman.2014.05.106.

[45] X. Ma, C. Zhang, K. Li, Hybrid modeling and efficiency analysis of the scroll compressor used in micro compressed air energy storage system. Appl. Therm. Eng. 161 (2019) 114139, https://doi.org/10.1016/j.applthermaleng.2019.114139.

[46] M. Heidari, M. Mortazavi, A. Rufer, Design, modeling and experimental validation of a novel finned reciprocating compressor for isothermal compressed air energy storage applications. Energy 140 (2017) 1252–1266, https://doi.org/10.1016/j.energy.2017.09.031.

[47] Kopernio, A Review of Thermal Energy Storage in Compressed Air Energy Storage System | Kopernio, https://kopernio.com/viewer?doi=10.1016/j.energy.2019.115993&route=6, 2019. (Accessed 20 December 2019).

[48] B. Zalba, J.M. Marin, L.F. Cabeza, H. Mehling, Review on thermal energy storage with phase change: materials, heat transfer analysis and applications. Appl. Therm. Eng. 23 (2003) 251–283, https://doi.org/10.1016/S1359-4311(02)00192-8.

[49] A.S. Alsagri, A. Arabkoohsar, H.R. Rahbari, A.A. Alrobaian, Partial load operation analysis of trigeneration subcooled compressed air energy storage system. J. Clean. Prod. (2019) 117948, https://doi.org/10.1016/j.jclepro.2019.117948.

[50] A. Arabkoohsar, Combined steam based high-temperature heat and power storage with an organic rankine cycle, an efficient mechanical electricity storage technology. J. Clean. Prod. (2019), https://doi.org/10.1016/j.jclepro.2019.119098.

[51] E. Barbour, D. Mignard, Y. Ding, Y. Li, Adiabatic compressed air Energy storage with packed bed thermal energy storage. Appl. Energy 155 (2015) 804–815, https://doi.org/10.1016/j.apenergy.2015.06.019.

[52] P. Wang, P. Zhao, W. Xu, J. Wang, Y. Dai, Performance analysis of a combined heat and compressed air energy storage system with packed bed unit and electrical heater. Appl. Therm. Eng. 162 (2019) 114321, https://doi.org/10.1016/j.applthermaleng.2019.114321.

[53] Kopernio, Advances and Prospects in Thermal Energy Storage: A Critical Review, https://kopernio.com/viewer?doi=10.1360/N972016-00663&token=Wzg2NTE5MCwiMTAuMTM2MC9OOTcyMDE2LTAwNjYzIl0.CZbRRH52GnlGTd_p-bjcGgvs2nw, 2019. (Accessed 22 December 2019).

[54] G.M. Crawley, Energy Storage. World Scientific, 2017, https://doi.org/10.1142/10420.

[55] EPRI Home, https://www.epri.com/#/?lang=en-US, 2020. (Accessed 30 December 2019).

[56] Enel In Italy—Company, https://corporate.enel.it/en/company/in-Italy. (Accessed 30 December 2019).

[57] P. Valdivia, R. Barraza, D. Saldivia, L. Gacitúa, A. Barrueto, D. Estay, Assessment of a compressed air Energy storage system using gas pipelines as storage devices in Chile. Renew. Energy 147 (2020) 1251–1265, https://doi.org/10.1016/j.renene.2019.09.019.

[58] Y. Li, L. Pan, K. Zhang, L. Hu, J. Wang, C. Guo, Numerical modeling study of a man-made low-permeability barrier for the compressed air energy storage in high-permeability aquifers. Appl. Energy 208 (2017) 820–833, https://doi.org/10.1016/j.apenergy.2017.09.065.

[59] A.J. Pimm, S.D. Garvey, M. de Jong, Design and testing of energy bags for underwater compressed air energy storage. Energy 66 (2014) 496–508, https://doi.org/10.1016/j.energy.2013.12.010.

[60] Lightsail. http://www.lightsail.com/cgi-sys/defaultwebpage.cgi.

[61] M. Cheayb, M. Marin Gallego, M. Tazerout, S. Poncet, Modelling and experimental validation of a small-scale trigenerative compressed air energy storage system. Appl. Energy 239 (2019) 1371–1384, https://doi.org/10.1016/j.apenergy.2019.01.222.

[62] S. Lv, W. He, A. Zhang, G. Li, B. Luo, X. Liu, Modelling and analysis of a novel compressed air energy storage system for trigeneration based on electrical energy peak load shifting. Energy Convers. Manag. 135 (2017) 394–401, https://doi.org/10.1016/j.enconman.2016.12.089.

[63] A.L. Facci, D. Sánchez, E. Jannelli, S. Ubertini, Trigenerative micro compressed air energy storage: concept and thermodynamic assessment. Appl. Energy 158 (2015) 243–254, https://doi.org/10.1016/j.apenergy.2015.08.026.

[64] M. Cheayb, M. Marin Gallego, S. Poncet, M. Tazerout, Micro-scale trigenerative compressed air energy storage system: modeling and parametric optimization study. J. Energy Storage 26 (2019), https://doi.org/10.1016/j.est.2019.100944.

[65] Y. Najjar, N. Jubeh, Comparison of performance of compressed-air energy-storage plant with compressed-air storage with humidification. Proc. Inst. Mech. Eng. A. J. Power Energy 220 (2006) 581–588, https://doi.org/10.1243/09576509JPE246.

[66] Baker Hughes, a GE Company, Liquid Air Energy Storage, https://www.bhge.com/industrial/energy-storage/liquid-air-energy-storage, 2019. (Accessed 30 December 2019).

[67] H. Guo, Y. Xu, H. Chen, X. Zhou, Thermodynamic characteristics of a novel supercritical compressed air energy storage system. Energy Convers. Manag. 115 (2016) 167–177, https://doi.org/10.1016/j.enconman.2016.01.051.

[68] M.J. Moran, H.N. Shapiro, D.D. Boettner, M.B. Bailey, Fundamentals of Engineering Thermodynamics, John Wiley & Sons, 2010.

[69] A. Arabkoohsar, G.B. Andresen, Thermodynamics and economic performance comparison of three high-temperature hot rock cavern based energy storage concepts. Energy 132 (2017), https://doi.org/10.1016/j.energy.2017.05.071.

[70] A. Arabkoohsar, H. Nami, Thermodynamic and economic analyses of a hybrid waste-driven CHP–ORC plant with exhaust heat recovery. Energy Convers. Manag. 187 (2019) 512–522, https://doi.org/10.1016/j.enconman.2019.03.027.

[71] F.M. White, Fluid Mechanics, McGraw-Hill, 1986.

[72] F.P. Incropera, T.L. Bergman, A.S. Lavine, D.P. DeWitt, Fundamentals of Heat and Mass Transfer. (2011), https://doi.org/10.1073/pnas.0703993104.

[73] A. Arabkoohsar, Non-uniform temperature district heating system with decentralized heat pumps and standalone storage tanks. Energy 170 (2019) 931–941, https://doi.org/10.1016/j.energy.2018.12.209.

[74] A. Arabkoohsar, Combination of air-based high-temperature heat and power storage system with an organic Rankine cycle for an improved electricity efficiency. Appl. Therm. Eng. 167 (2020) 114762, https://doi.org/10.1016/j.applthermaleng.2019.114762.

[75] M. Farzaneh-Gord, A. Arabkoohsar, M.D. Dasht-bayaz, L. Machadob, R.N.N. Koury, Energy and exergy analysis of natural gas pressure reduction points equipped with solar heat and controllable heaters, Renew. Energy 72 (2014) 258–270.

CHAPTER FOUR

Pumped hydropower storage

Ahmad Arabkoohsar[a] and Hossein Namib[b]

[a]Department of Energy Technology, Aalborg University, Esbjerg, Denmark
[b]Department of Mechanical Engineering, Faculty of Engineering, University of Maragheh, Maragheh, Iran

Abstract

Pumped hydropower storage (PHS), also known as pumped-storage hydropower (PSH) and pumped hydropower energy storage (PHES), is a source-driven plant to store electricity, mainly with the aim of load balancing. During off-peak periods and times of high production at renewable power plants, low-cost electricity is consumed to pump water to a high elevation reservoir. In this way, the surplus electrical power is stored in the form of gravitational potential energy. When electricity demand increases, the stored water is released to drive the employed hydraulic turbine(s) of the system and actuate a coupled electricity generator to produce power. The outlet flow from the higher reservoir can be controlled to provide variable output power. The roundtrip efficiency of a PHS plant can reach up to 85%, which is the highest percentage among mechanical energy storage (MES) technologies. Also, the capacity of such plants can be extremely large, up to a few thousand megawatts. The main disadvantages of PHS systems are the limitations in water availability and topography challenges as well as high capital cost. Further, appropriate sites for this technology seem to be available in the natural environment and therefore there are also ecological and social concerns to overcome. Even considering these challenges and drawbacks, PHS is by far the most widely implemented energy storage technology in the world due to the previously mentioned advantages and its considerable economic benefits via facilitating the supply of cheap electricity at expensive times and spot prices and high efficiency. This chapter gives information about the fundamentals of PHS systems, a history of different kinds of PHS systems, and the state of the art of the technology. Then, it presents the basic mathematical model required for analyzing a PHS facility followed by a discussion on the future perspectives of this technology.

4.1 Fundamentals

In previous chapters, we addressed the necessity of energy storage systems, specifically electricity storage systems. We argued that energy storage units are highly beneficial technically and economically not only for renewable power plants but also for conventional power plants and grids [1]. Regarding renewable plants, it is clear that electricity storage systems can make fluctuating power output dispatchable, which is extremely important for keeping the stability of the grid [2]. In conventional systems, on the other hand, energy storage units can increase the economic benefit via facilitating proper timing of power sales to the market and giving them strong flexibility [3]. For instance, consider an electricity grid with high penetration of conventional power plants including nuclear plants, Rankine steam power plants, and others. In this system, when the demand increases or a production unit suddenly runs into a technical problem, the frequency of the grid falls and, naturally, the outlet power of these plants cannot be shifted up and down as quickly as needed. Hence, employing an energy storage unit can maintain the required balance between the electricity supplier systems and the demand, eliminating the need for expensive auxiliary services [4].

Among the several existing and developing MES systems, the pumped hydropower storage (PHS) system is the most mature technology. This has made it the most broadly implemented energy storage system worldwide. By 2017, PHS covered more than 95% of the total in-service electricity storage capacity of the globe with a total capacity of about 184 GW [5]. Until 2011, around 40 PHS facilities were operating globally (mostly in the United States), while just two compressed air energy storage facilities (one in Germany and one in the United States) were in operation [6]. PHS systems can be divided into two main configurations: open-loop and closed-loop systems. In the former, the system is coupled to a natural water system such as a river (the lower reservoir is a dam connected to a natural water source), while in the latter the reservoirs are not linked to any natural body of water and it is only the upper and lower reservoirs between which the mechanical components of the system are placed and work. Fig. 4.1 illustrates the schematic diagram of a simplified PHS system in a closed-loop configuration (the lower reservoir is not connected to any river or natural water stream).

The operating principle of a PHS system is quite simple. As shown in Fig. 4.1, a PHS consists of a few main components, namely, an upper reservoir, a hydraulic turbine/pump that is connected to a motor/generator, a lower reservoir, the flow canal/piping system, and control systems (e.g., flow control valve). In the charging mode, the surplus power that is to be stored is used by the motor to drive the turbine/pump, which works on its pump mode at this stage. The pump transfers water from the lower reservoir to the upper reservoir to store the surplus electricity in the form of increased potential energy in the system. In the discharging mode, the direction of water flow and operation of components is reversed. The control valve opens, depending on the level of power needed to be produced, and water from the upper reservoir at high pressure flows through the hydraulic turbine/pump (which is in

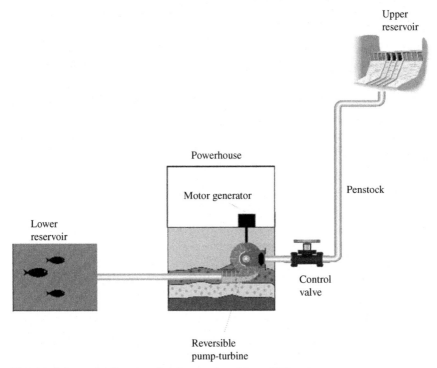

Upper
reservoir

Powerhouse

Motor generator

Penstock

Lower
reservoir

Control
valve

Reversible
pump-turbine

Fig. 4.1 Schematic diagram of a simple closed-loop PHS system.

turbine operation mode) and produces rotational work that is fed to drive the generator. Finally, the discharged water at much lower pressure lies within the lower reservoir [7].

In the following, we present an overview of the technologies used in the past and to be used in the future for each of the main components of a PHS system. During the discharge process, the employed turbine converts hydraulic power in water flow to mechanical power and acts as the main unit of the system. To provide an efficient and cost–effective energy storage system, advances in designing this component are required to improve the total performance of the plant [8]. The selection of the turbine to be employed in PHS systems is a function of cost estimates and operational parameters. There is a significant diversity of designs for different operating conditions. Nevertheless, proper turbines can be listed in four different categories: impulse, reaction, pump hydro, and gravity turbines [9].

Impulse turbines: These types of turbines normally utilize the water velocity to move the runner while discharging to around atmospheric pressure. An impulse turbine is appropriate for low–flow and high–head applications. Pelton, Turgo, and cross-flow turbines are the three main kinds of impulse turbines.

Lester Allan Pelton patented the Pelton turbine in 1880. This hydraulic turbine was invented to be used for hydroelectric generation in geographical conditions where high headwater with a small flow is available [10]. A laboratory-scaled Pelton turbine, which was studied by Agar and Rasi [10], is illustrated in Fig. 4.2. As shown, evenly spaced vanes fit the wheel. Against the traditional wheels, these vanes consist of two cups. A water jet hits these cups by one or more nozzles to convert the water's kinetic energy into impulses.

The Turgo turbine, a modification of the Pelton turbine, was designed by Gilkes in 1919 for medium-head applications. Since its invention, a large number of facilities around the world have been equipped with this turbine. Normally, Turgo turbines operate with an efficiency of around 87%. This kind of impulse turbine not only increases the capacity of hydropower generation but also contains a nozzle-and-spear injector system, which is the main characteristic of the Pelton turbine for flow adjustment [11]. Also, Turgo turbines have a greater specific speed and can handle a greater volume of water compared to Pelton turbines of the same diameter, leading to lower installation costs [12]. The Turgo turbine was developed to compensate for the main problems of the Pelton turbine. In contrast to the Pelton turbine, the water jets in the Turgo expanders are directed to a specific sharp angle in contradiction to the runner rotation plane, while the water releases from the opposite side of the runner. In this way, the interference of the outflow with the jets and runner (an existing problem in the Pelton design) can be minimized. In addition to this, water jets hit more than one blade at a time and each blade has a complex 3D surface. Therefore, fast and complete discharge

Fig. 4.2 Schematic diagram of a Pelton turbine [10].

Fig. 4.3 A typical Turgo turbine; front-inlet side *(left)*, back-outlet side *(right)* [11].

of water can be achieved to minimize the outlet water energy [13]. Fig. 4.3 shows a typical Turgo turbine.

A cross-flow turbine is a drum-shaped machine known as a Bánki-Michell or Ossberger turbine. In contrast to most water turbines, which have radial or axial flows, in a cross-flow turbine the water flow across the turbine blades passes through the turbine transversely [14]. This kind of impulse turbine operates at lower speeds and is appropriate for low-head and high-flow streams [15]. This machine was developed to deal with a larger quantity of water flows and lower heads compared with the Pelton turbine. Ebrahimi et al. [16] developed an experiment to study the effect of a cross-flow turbine on an erodible bed with the aim of maintenance of this kind of turbine with constant foundations. Fig. 4.4 illustrates a simplified diagram of a cross-flow turbine. The cross-flow turbine is shown with both horizontal and vertical inflows.

Reaction turbines: A reaction turbine produces electricity utilizing mutual action of pressure and moving water [17]. Compared with impulse turbines, reaction turbines don't change the water flow direction too drastically. In addition to this, reaction turbines are generally employed in facilities with higher flows and lower heads compared with impulse turbines. Bhatia

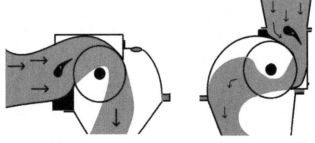

Fig. 4.4 Cross-flow turbine; with a horizontal inflow *(left)*, with a vertical inflow *(right)*.

[18] stated that "a reaction turbine is a horizontal or vertical wheel that operates with the wheel completely submerged, a feature which reduces turbulence." The operation of this kind of turbine is attained when the rotor is filled with water and is enclosed in a pressure casing. A draft tube act as a diffuser to discharge the water and exists below the runner in all reaction turbines. In this way, the effective head will be increased due to a reduction in the static pressure below the runner [19]. Theoretically, water is under pressure at the central point of a reaction turbine and causes rotation via escaping from the blades' ends. Reaction turbines can be divided into two main groups of Francis and propeller turbines [17].

The Francis turbine is the most common water turbine in use. It was developed by James Bicheno Francis during the 1850s [20]. A change in the water direction passing through the machine is the key feature of this turbine. The Francis turbine is designed to be extremely flexible and can be used in facilities with different flow rates and heads [21]. Fig. 4.5 shows a Francis turbine coupled with a generator as well as a simple Francis turbine [22]. A Francis turbine can be equipped with a radial flow runner or a mixed axial/radial flow one. A draft tube and wicket gates are the next main components of this kind of turbine.

Propeller turbines are axial turbines that have three to six blades on their runners. Normally, the number of blades on the axial runner depends on the water head and turbine design [17, 23]. This kind of turbine is an appropriate expander for high-flow and low-head cases and is well suited for small and mini-hydro Schemes [24, 25]. However, propeller turbines for micro-hydro

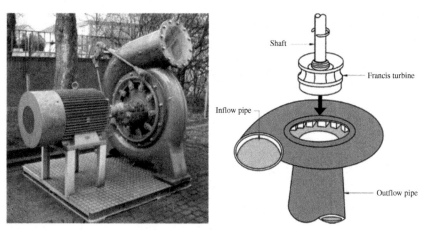

Fig. 4.5 A Frances turbine coupled with a generator [22] *(left)*, a simple Francis turbine [17] *(right)*.

facilities are still in the early stages. Kaplan, tube, Straflo, and bulb turbines are different kinds of propeller turbines. Among these, the Kaplan turbine is the most utilized and is equipped with adjustable blades [26]. A typical Kaplan turbine is shown in Fig. 4.6.

Pump hydro turbine: Pump hydro turbines or reversible pump turbines are naturally centrifugal machines and can be utilized as pumps during the charging process to transfer water into the higher reservoir and at the same time can operate as turbines during peak demand to generate power [27]. Different configurations of pump hydro turbines exist including single or multiple stage, horizontal or vertical, fixed or variable speed, combined or individual actuation of guide vanes, and with or without cylindrical ring gates [28]. The ability of rapid change in the operation mode makes a pump turbine an appropriate machine for electricity generation and storage. This is of great importance for handling electrical grid fluctuations. In addition, being compact and having multiple stages limits the application of pump turbines in freshwater with low values of solid contents. Francis-type pump turbines are usually custom-designed to satisfy the demand-side require-ments in storage applications.

Archimedes (Gravity) pump turbine: The screw pump is the oldest positive displacement pump [29]. Utilizing Archimedes screws as a pump has a long history and has recently been proposed for use as a turbine. Using the screws as a pump, water will be transferred to the upper dam by turning a screw-shaped surface inside a pipe. This hydro machine was mainly utilized to transfer relatively large volumes of water at low heads [30]. Also, this device completely satisfies application operators because it is extremely reliable and

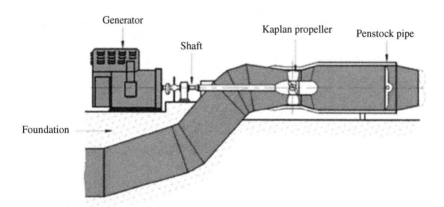

Fig. 4.6 A typical Kaplan turbine [17].

Table 4.1 Classification of turbines with their appropriate application heads [30].

Head meter (m)	Impulse turbine	Reaction turbine	Gravity turbine
Low (<10 m)	Cross-flow	Propeller Kaplan Francis Alden Bulb Straflo	Overshot waterwheel Pitchback waterwheel Breastshot waterwheel Archimedes screw
Medium (10–50 m)	Cross-flow Turgo Multi-jet Pelton	Francis	–
High (>50 m)	Turgo Multi-jet Pelton	Francis	–

durable [31, 32]. Although the gravity turbine operates like an overshot waterwheel, the clever shape of the helix allows for faster rotating resulting in greater efficiency of power conversion (more than 80%) [33, 34].

An overview of the different turbines employed in PHS plants: Table 4.1 presents a summary of the different turbines. Suitable machines for different head levels are suggested to give a clear roadmap for readers and designers of PHS systems regarding the selection of proper turbines. Note that some of the turbines mentioned in this table are not discussed in the preceding sections.

Generally, the performance of a PHS facility is a function of the efficiency of the employed turbines (and the pumps), which is, of course, a function of the plant capacity as well [35]. The efficiency of some different typical hydraulic turbines with a change in water flow rate is shown in Fig. 4.7 [36]. Four water turbines with a capacity of 3, 5, 7, and 9 MW are considered in this plot. As shown, in all cases, turbine efficiency hits its maximum value in a specific volumetric flow rate. The greater the capacity of the turbines, the greater the optimal water flow rate. According to the figure, the smaller turbines reach maximum efficiency faster. For these typical cases, the optimum flow rates are about 10, 16, 22, and 30 m^3/s for the 3, 5, 7, and 9 MW turbines, respectively [36]. Overall, maximum efficiency of the turbines can reach up to 85%.

Like turbines, the efficiency of the pump is critical in the total efficiency expected from a PHS system. As discussed, pumps operating at a variable speed perform more efficiently than pumps at constant speed [37]. Fig. 4.8 is an efficiency diagram for different typical variable-speed pumps. This figure gives the value of the pumps' efficiencies versus the consumed

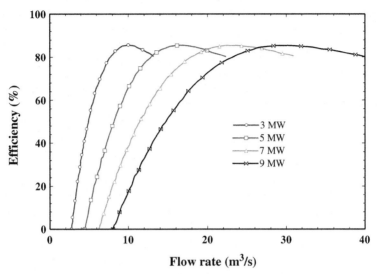

Fig. 4.7 Turbine efficiency versus water flow rate [36].

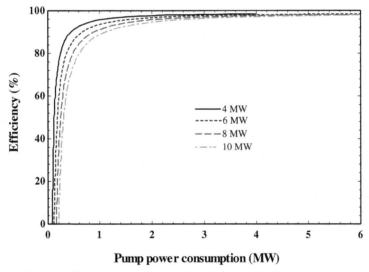

Fig. 4.8 Efficiency of variable-speed pumps versus consumed power.

power [36]. Expectedly, the maximum efficiency of pumps is much greater than that of hydraulic turbines. For pumps, efficiency can reach up to 95%. Here, typical variable-speed pumps with capacities of 4, 6, 8, and 10 MW are shown. In all the cases, especially those with a capacity greater than 4 MW, variable-speed pumps behave like each other with almost the same efficiency. Furthermore, the initial power of the 4- and 10-MW variable-speed pump turbines is 0.08 and 0.2, respectively [36].

On the other hand, motor-generators are special devices that can operate as both motors and generators depending on the operation mode. Naturally, in the pump mode, a motor-generator converts electrical power into rotational power and vice versa. In PHS facilities, motor-generators are employed to drive the installed pump turbines in the charging mode (motor operation mode driving the pump) and generate electricity (in the discharging mode being driven by the turbine). Motor-generators have been in use for many years. However, low efficiency and high costs made it difficult for them to compete with modern welding power units and thus these devices are no longer manufactured [38]. Motor-generators developed for storage applications may reach a capacity of 360 MVA with rotating in one or two directions available in both variable and constant speeds [39]. In addition, this device can be coupled with all unit arrangements and configurations [30]. In addition to cost, factors such as physical size, unit size, speed, cooling, and reliability should be considered when selecting a motor-generator [40]. Variable-speed motor-generators were introduced in 1977 and have been utilized in numerous storage facilities [41]. These devices can change the rotation speed of a normally fixed-speed motor. In addition to energy storage facilities, air conditioning systems are also equipped with these devices. Compared with constant-speed drivers, variable-speed motor-generators have several advantages [42], namely, efficient operation, reduced noise generation at part-load operation, and reduced wear on mechanical elements like belts and bearings. In addition, variable-speed motor-generators are also utilized to control pumps on variable-flow pumping systems. This is another advantage of variable-speed motor-generators compared with fixed-speed pumps. As such, an adjustable-speed pump is more efficient than a fixed-speed pump and thus it is a more feasible solution [37].

4.2 State-of-the-art and practice

The first PHS facility started to operate in Switzerland in the 1890s. It was designed based on the individual pump and turbine units. Thereafter, the implementation of PHS systems continued at a very slow or at a very fast pace depending on the times. For example, after the appearance of an integrated pump turbine for PHS systems in the 1950s (separated pump and turbine designs were the only available solution at the time) and the emergence of nuclear power plants that needed PHS systems to cover the peak demand of grids, the pace of deployment of PHS sites grew significantly [43]. Before this, PHS systems were not taken seriously. In the 1990s, when the price of natural gas dropped considerably and made conventional fuel-driven power plants much more cost-effective, the deployment of PHS systems slowed down. Another reason for such slow growth is environmental concerns associated with this technology. For example, for peak-shaving of the New York City power grid in 1963, a PHS plant was proposed to be constructed at Storm King Mountain. It was to be the world's largest hydropower storage project at that time, however, environmental protection groups fought against it. They thought the project posed a serious threat to the Hudson River, fisheries, the local water supply, and the scenic beauty. Therefore, the project was stopped and eventually terminated [6]. Another example of discord between environmental groups and project developers is the Richard B. Russell Dam. In this case, however, despite initial opposition, a good collaboration between stakeholders and assessment of environmental effects led to eventual approval of the project. The Russell hydropower plant was finalized and started to operate in 1986, while required permissions were gained in 2002 to commercial operation of four employed pumps and in this way the PHS plant was completed [6]. Fig. 4.9

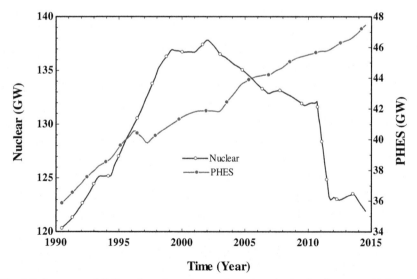

Fig. 4.9 Progress of PHS and nuclear power plant capacity in the European Union.

shows the variation in the capacity of PHS units and nuclear power plants from 1990 to 2015 in the European Union [44].

The major growth of global PHS capacity, especially in recent decades, is associated with universal awareness about the need for increasing the share of renewable energy systems. Thus, a wide spectrum of renewables, mainly wind and solar power plants, highlights the importance of PHS facilities. Fig. 4.10 illustrates the development of PHS technology versus the profile of the increase in the capacity of solar and wind power plants [44]. According to the figure, there is great compatibility with the increase of PHS systems with solar and wind power plants. For these power plants, the fluctuations of the energy source can be compensated by the PHS unit storing the surplus energy of the plants at off-peak times and compensating for them at peak demand periods. Several studies report that growing electricity supply via wind resources has motivated the development of PHS facilities [45–47].

In the sections that follow, we review some of most recent studies on PHS systems that have led to significant steps forward in not only the state of the art but also the state of practice of this technology in different national energy systems around the world. This research focuses mainly on three themes: "innovations and advancements in PHS systems," "integration with renewable-based power plants," and "case studies of PHS systems." Since the number of works on each of these matters is in the thousands, we review only some of the most recent works here.

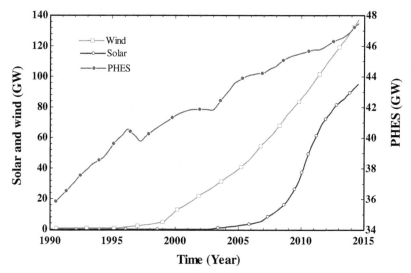

Fig. 4.10 Progress of PHS, solar and wind capacities in EU28.

Innovations and advancements in PHS systems: This category of works includes improvements in design, components, operation, and control of PHS systems. For example, one of the most important focuses not only for PHS systems but also for any other energy system is digitalization to increase the flexibility of the system (via advanced grid-supporting services and providing storage capacity), especially for the sake of an optimal integration with renewable power plants. This will cause fundamental changes in the conventional methods for the design, development, upgrade, operation, and maintenance of PHS systems [48]. Another important example is the use of variable-speed turbomachinery in PHS systems, which was previously discussed. Variable-speed turbomachinery in PHS technologies is already quite mature in terms of development, however, the application of these machines in real-life PHS systems is not common as of yet. Some of the main reasons for this slow progress might be the very long time required for an applied change due to the complex administrative procedures in this area, the greater initial costs required for variable-speed machines (7%–15% of the total cost of the plant [49]), and lack of concrete confidence regarding the cost-effectiveness of PHS systems in many countries as well as environmental concerns [50]. Iliev et al. [51] explained how a variable-speed operation could be managed for Francis turbines, how variable-speed turbines might enhance the off-design operation efficiency of PHS systems by up to 10% for larger head variations, and how they can help the storage unit quickly, accurately, and smoothly respond to grid load fluctuations. The advantages of adjustable-speed PHS systems are assessed by Yang et al. [52] to regulate wind power variations considering physical qualities and economic indicators. Vargas-Serrano et al. [53] studied the economic benefit of employing a variable-speed machine instead of a fixed-speed one to enhance the flexibility of the plant and allow for the provision of additional services in pump mode. The Grimsel 2 energy storage plant, located in Switzerland, was used as the case study. It was revealed that improving the fixed-speed plant to a variable-speed plant increased the total revenue of the plant by up to 58%. Haney et al. [54] estimated the required capital cost of a permanent magnet synchronous generator used for low-volume variable-speed PHS manufacturing.

Menendez et al. [9] investigated the effects of air pressure on Francis turbine performance employed in a PHS plant equipped with an underground reservoir. The main aim was to show the effects of air pressure change during the design of PHS systems. Differences between variable- and fixed-speed storage plants used in a security-constrained unit commitment were studied by Salimi et al. [55] according to the computation of added values in the electricity market. The main aim of this research was to provide a security-based coordination approach for wind power plants and PHS systems in the unit commitment problem. It was reported that utilizing adjustable-speed energy storage makes wind power plants more effective. A novel solution to obtain roundtrip efficiency of PHS plants with underground water reservoirs was presented by Menéndez et al. [56]. CFD simulation and analytical modeling were used to assess the performance of a hydropower plant with 214.7 and 124.9 MW Francis turbines and power consumption of 199.7 and 114.8 MW. It was revealed that unlike in conventional plants, the roundtrip efficiency of these kinds of PHS facilities is a function of underground reservoir pressure. As an illustration, reducing the reservoir pressure to -100 kPa led to a reduction in roundtrip efficiency from 77.3% to 73.8%. The effects of groundwater exchanges on the performance of PHS systems were analyzed by Pujades et al. [57]. Underground PHS plants have an interaction with the surrounding porous medium via groundwater exchanges and these exchanges affect the employed pump and turbine efficiencies.

To drive optimal performance guidelines of PHS plants, Zhao et al. [58] proposed a new optimization framework. This framework was divided into three correlated stages: nonlinear modeling, optimization of strategy, and how to make a decision. Tian et al. [59] did a risk assessment of PHS systems using downside risk constraints to find a zero-risk operation and bidding strategies (for selling electricity and buying electricity to/from the grid) for the systems. Ak et al. [60] studied a cascade PHS plant from the point of view of monetary benefits. The main aim of this study was to investigate operating strategies for cascade PHS plants consisting of already existing hydropower facilities. As an example, revenue gain from the cascade multi-reservoir facility located in the Coruh Basin of Turkey (total 1100 MW) was estimated to be between 5 and 19.2 million €. Cheng et al. [61] developed a novel model based on mixed-integer linear programming to find the optimal performance of PHS plants in each hour serving several locally available power grids instead of just one. The main objective of this study was to reduce the differences of peak-valley for residual load series. A comparative study was done by Connolly et al. [62] to compare three different applied operation approaches. A 2-GWh storage plant with a 360-MW turbine and 300-MW pump was considered to obtain optimal profit implying cost arbitrages on 13 various electricity markets. It was shown that operating the energy storage facility under optimized conditions via day-ahead real or exact electricity costs led to obtaining 97% of the entire profits. Otherwise, the feasibility of the plant would decline considerably.

Wind and Solar plants integration: Jurasz et al. [63] studied the connection of hybrid power plants to the power system consisting of solar and wind energy and equipped with a PHS system. It was assumed that the PHS system can balance the varying power output from the wind turbine and photovoltaic (PV) plants. Daneshvar et al. [64] proposed a two-stage stochastic model for optimal scheduling of a combined energy system including a PHS unit equipped with wind-thermal power plant, while competitive interactions between power generation elements were taken into account. Within the first stage, more attention was paid to the day-ahead programming of the system. The second stage was focused on balancing market dispatch. A parametric study was also carried out to investigate the effects of the design parameters on system operation. The results showed that decreasing the ramp up/down factors and capacity of the transmission line, and increasing the minimum up/down restrictions, led to reduced system flexibility and increased total energy cost. Xu et al. [65] investigated the performance of a PHS plant combined with solar-wind energy sources from the stability point of view. A unified model was proposed and a parametric uncertainty was quantized on the outage contribution of the plant. The feasibility of employing the PHS plant in the combined power systems was shown in this research under both steady and fault scenarios. Application of PHS technology in power transmission from wind farms was studied by Su et al. [66]. In addition, hydraulic constraints of the PHS facilities and errors associated with wind power estimation were taken into account. Also, to investigate the effects of key parameters on the overall performance of the combined system, a sensitivity analysis was carried out. Results showed that integrated transmission of wind energy and PHS not only increases profit significantly but also reduces the negative effect of wind power fluctuations efficiently. Wang et al. [67] proposed a feasibility study of approaching the potential of small PHS facilities in hybrid multi-energy (wind and PV) applications. This study was aimed at minimizing output fluctuations of wind and solar power, while it was constrained by the programmed electricity generation and other common limitations. Dujardin et al. [68] considered the application of PHS to balance the intermittency between different renewables, showing that up to 25% enhancement is needed in seasonal storage for a fully balanced renewable system in Switzerland. It was also demonstrated that a PV contribution of less than 20% is required for the best performance of the combined system, including wind, PV, and PHS plants. Pérez-Díaz et al. [69] analyzed the contribution of PHS plants to decreased scheduling costs of isolated power stations, with special attention on integrated wind power. Results indicated that this energy storage technology decreases scheduling costs by 2.5%–11%.

A linear programming method was employed by Brown et al. [70] to reach the maximum derivation of the wind potential of an island via optimizing the size of the reservoirs and pumps. The benefits of optimal combination of the Lake Turkana Wind Power Project in Kenya with a PHS plant were studied by Murage et al. [71]. Maximizing the expected revenue over the planning horizon was also considered in this study. It was revealed that since the daily pattern of the wind speed does not match the daily load pattern, employing the PHS plant decreased the entire system's electricity production shortage by 46%. Kapsali et al. [72] investigated the economic feasibility of integrating wind power and a PHS plant to supply electricity demand to an island in the Aegean Sea with special attention on peak demand. To obtain the optimal possible configuration, the net present value of the project was maximized. Results revealed that excellent economic and technical performance was achievable and the contribution of renewables in the energy market of the island was doubled. A wind-powered PHS plant was introduced by Katsaprakakis et al. [73] for isolated insular power systems of two islands (Karpathos and Kasos) in the Agean Sea without any connection to the main power grid. The main aim was to increase the share of wind power instead of fossil fuel consumption in the thermal power plant. Results proved that the project is economically feasible and the payback period for the invested capital cost was estimated to be around 5–6 years. A standalone PV power-producing system equipped with a PHES plant was modeled and economically optimized by Ma et al. using a genetic algorithm [74]. Maximizing the reliability of the power supply and minimizing lifecycle costs were supposed to be the objective functions. Results revealed that this method of analyzing and optimizing seemed to be effective and can be applied to assess other similar cases. Javanbakht et al. [75] studied the transient performance of a hybrid PV-PHES system to observe system performance during energy storage (charging) and supplying (discharging). Combining the PHES facility with PV systems to propose a sustainable and continuously working power plant was studied theoretically by Margeta et al. [76]. Manolakos et al. [77] investigated the application of a PHES plant to supply the energy demand (electricity consumption associated with lighting, TVs, and refrigerators) of 13 houses in a remote village via a standalone PV including 300 PV modules. The employed PHES was comprised of two 150 m³ water reservoirs and a generator with 7.5 capacity.

A few samples of research works on the PHS systems in different case studies: Melikoglu [78] analyzed the global development of PHS systems and Turkey's potential according to Vision 2023 hydropower wind and solar energy targets. Since Turkey's government plans to increase hydropower capacity up to

36 GW, PHS systems should be the center of attention. This is not the end of the story, and increasing solar and wind energy capacities up to 3 and 20 GW makes the development of PHS systems a hot topic in Turkey. Xu et al. [36] studied the potential of application of a combined energy system including solar, wind, and PHS installations with a real situation in Xiaojin, China. The PHS system was used to store the surplus wind and solar power and generate electricity from a local river water flow. Winde et al. [79] studied the possibility of utilizing gold mines located in South Africa (Far West Rand goldfield) as the lower reservoir of PHS systems. The fact that the locally available gold mines were suffering from frequent power outages highlighted the importance of utilizing energy storage technology in this area. Karimi et al. [80] proposed a method to obtain the Siahbishe PHS facility (the first PHS project in Iran began in the 1970s) in the Iran power grid using actual data. It was estimated that the integration of this PHS plant to the main grid has an annual economic benefit of approximately 94 million dollars. Authors of Ref. [81] focused on the importance of ternary PHS plants as an effective parameter on the frequency response of the United States. Ternary PHS plants are interesting due to their fast response during peak demand. To show the feasibility of employing ternary PHS systems, the frequency response of the Western Interconnection was compared with and without this technology. Lu and Wang [82] assessed the potential of Tibet for PHS deployment using a geographic information system and concluded that there is a huge potential for such power storage plants. They mapped the appropriate sites in Tibet and did an assessment of them based on the distances of the sites to the electricity grids. Gutierrez and Arantegui [83] accomplished an assessment for new PHS systems in Europe and concluded that there is a practical storage capacity of up to 29 TWh for only two existing reservoirs and considering all the social and environmental concerns. This capacity is much larger than what exists in operation at the moment. Kusakana [84] presented a study aimed at finding the optimal operation strategy of PHS systems (if being built) in the electricity market of South Africa. This work developed a mathematical model that describes the optimal operation scheduling for an arbitrary PHS system to maximize the economic benefit of such an energy storage unit in this case study market. Ko et al. [85] presented an economic analysis of PHS systems for increasing the share of renewable energy in South Korea. They concluded that PHS systems are perfectly feasible in this energy system and could significantly contribute to the targeted renewable energy expansion of the Korean government by providing dispatchable power for the renewable plants of the country and their peak-shaving services. Rehman et al. [43] present a table showing the total PHS capacity of different countries of the world.

A methodology to obtain the appropriate size PHES plant to exploit surplus local wind power was applied for the Lesbos wind farm in Greece by Kaldellis et al. [86]. They evaluated the system operation for a year with hourly reportable data and calculated energy losses. Application of PHES plants to harvest the rejected power from wind farms was investigated by Kapsali et al. [87] who also carried out a sensitivity analysis considering economic indices. The optimal scheme of pumped storage was examined to integrate an existing large-scaled PHES plant with a new pumping station unit. The performance of a PHES plant was assessed in a traditional hydroelectric power plant by Anagnostopoulos et al. [88]. The system was examined for 1 year to obtain the inflow water change. In addition, an economic analysis was carried out based on the financial conditions of Greece. Karimi Varkani et al. [89] proposed a novel self-scheduling approach based on stochastic programming methods for the combined procedure of wind and PHES plants in power markets. The uncertainty of wind-power generating was also modeled via a neural network-based procedure.

4.3 Mathematical model

A typical PHS system consists of lower and upper reservoirs, a variable-speed pump, water turbine, generator, and pipelines. In this section, we present the mathematical modeling of different components and units of a typical PHS system. The mathematical model for different parts of the PHS system is presented in a way that all the needed parameters can be easily taken out from the technical manuals. Therefore, without any considerable difficulty, readers can use the presented model for analyzing the performance of a PHS system.

4.3.1 Pipelines

Governing equations associated with the pipelines are presented here. Continuity and momentum equations are the fundamental equations for one-dimensional simulation of pipelines and can be formulated generally as [90]:

$$V\frac{\partial H}{\partial x} + \frac{\partial H}{\partial t} + \frac{a^2}{g}\frac{\partial V}{\partial x} + \frac{a^2 V}{gA}\frac{\partial A}{\partial x} = \sin\theta . V \qquad (4.1)$$

$$g\frac{\partial H}{\partial x} + V\frac{\partial V}{\partial x} + \frac{\partial V}{\partial t} + f\frac{V|V|}{2D} = 0 \qquad (4.2)$$

In these equations, V, H, α, A, θ, f, and D refer to the average flow velocity of pipeline section, piezometric water head in the pipeline, velocity of pressure wave, cross-sectional area of pipeline, the angle between the axis of pipeline and horizontal plane, Darcy-Weisbach coefficient of friction resistance, and the inner diameter of the pipe, respectively.

4.3.2 Water reservoir

It is supposed that natural water flows fill the upper reservoir, and in the charging mode, the surplus electricity produced by the local renewables will be utilized by the system's pumps to transfer water from the lower reservoir to the upper reservoir [36]. As a simplification, it is assumed here that the upper reservoir has a cubic shape. Thus, the hydropower station capacity can be written as [63]:

$$V_M = abh \tag{4.3}$$

In which a, b, and h are the length, width, and height of the reservoir, respectively. The volume of the water stored in the upper reservoir (V) can be expressed as:

$$V = V' + Q_{nf} + Q_{T/P} \tag{4.4}$$

In this equation, V' is the amount of water available in the reservoir from the previous hour and Q_{nf} is the supplied water via natural flow. $Q_{T/P}$ is positive when the pump is under operation (during the charging process) and is negative when the turbine is under operation (during the discharge process).

Evaporation is one of the loss sources for the amount of water in the reservoir assuming no seepage. Evaporated water volume can be estimated via the following equation [91]:

$$V_{eva} = \frac{ET_0}{3.6 \times 10^6} A \Delta t \tag{4.5}$$

Where ET_0 is the reference evapotranspiration level (mm/h), A is the reservoir area (m^2), and Δt is the time interval (s). The reference evapotranspiration level can be expressed as:

$$ET_0 = \frac{0.408\Delta(R_n - G) + \lambda \dfrac{37}{T_h + 273} u_{air}(e_0 - e_a)}{\Delta + \lambda(1 + 0.34 u_{air})} \tag{4.6}$$

In this equation, Δ refers to the saturation slope vapor pressure curve at T_h (kPa/°C), R_n is the net radiation (MJ/m^2-h), G is the density of soil heat flux

$(\mathrm{MJ/m^2\text{-}h})$, λ denotes psychometric constant $(\mathrm{kPa/°C})$, u_{air} is the air velocity 2 m above the reservoir surface (m/s), e_o is the saturation vapor pressure at environment temperature (kPa), e_a is the average hourly actual vapor pressure (kPa), and T_h is the hourly average environment temperature.

Precipitation is the next key parameter affecting the volume of reservoir water. Since upper reservoirs are open storage, any precipitation will add water to these reservoirs. As an example, in humid rainforests the average rainfall is reported to be between 1.5 and 4 m per year [91]. Then, it seems ignoring the added water via precipitation increases the error in the calculation of the volume of the stored water in the reservoir significantly. On the other hand, energy modeling and management of PHS plants greatly depends on the volume of the stored water. Therefore, more attention should be paid to obtain the exact value of available water in the reservoir. Amount of added water due to precipitation can be calculated from:

$$V_{pre} = \frac{I}{3.6 \times 10^6} A \Delta t \qquad (4.7)$$

Where I is the precipitation rate in mm/h. As such, Eq. (4.2) can be upgraded to the following general equation in which the losses are taken into account:

$$V = V' + Q_{nf} + Q_{T/P} - V_{eva} + V_{pre} \qquad (4.8)$$

4.3.3 Hydro turbine and pump

To design a PHS system, the employed turbine and variable-speed pump can be accounted as the main components. Much attention should be paid to obtain the proper sizing of these components. Undersized components lead to load losses and cause plenty of wasted energy. On the other hand, oversized components increase the initial investment cost and reduce efficiency. The models of the variable-speed pump and turbine are described below.

The volume of water pumped to the upper reservoir can be expressed as [63]:

$$E_P = \min\left(\frac{V_M - V'}{3600}, Q_P\right) \frac{\rho g}{\eta_P}\left(\frac{V'}{ab} + h'\right) \qquad (4.9)$$

$$Q_{Pump} = \frac{\eta_P E_P}{\rho g\left(\dfrac{V'}{ab} + h'\right)} \qquad (4.10)$$

Where E_p calculates the consumed power by the pump (mainly surplus energy by the locally available renewables) at every hour (kWh) to pressurize and supply the water Q_{pump} to the upper reservoir. η_P is the overall efficiency of the variable-speed pump (see Fig. 4.5). ρ, g, and h' are the water density, gravitational acceleration, and basic hydropower height, respectively. Q_p is the volume of pump throughput.

The quantity of the discharged water back to the lower reservoir during the discharge process can be written as:

$$E_T = \min\left(\frac{V'}{3600}, Q_T\right)\eta_T \rho g\left(\frac{V'}{ab} + h'\right) \tag{4.11}$$

$$Q_{dis} = \frac{E_T}{\eta_T \rho g\left(\dfrac{V'}{ab} + h'\right)} \tag{4.12}$$

Here, E_T (kWh) is power production by the employed turbine at each hour, while the volume of released water is Q_{dis}. Q_T is the water turbine throughput and η_T is the turbine efficiency. More details regarding the mathematical model of the turbines can be found in Ref. [92].

4.3.4 Generator

The first-order swing equation is developed for the generator, and three different operating conditions are considered for that [92]. The most general form is adopted for the single-machine isolated operation and can be formulated as:

$$J\frac{\pi}{30}\frac{dn}{dt} = M_t - M_g - \frac{30 e_g p_r}{n_r^2 \pi}\Delta n \tag{4.13}$$

For the case of operating with constant rotation speed, the following equation should be formulated:

$$n = n_c\left(f_g = f_c\right) \tag{4.14}$$

M_g and e_g values of zero result in off-grid operation and the governing equation can be expressed as:

$$J\frac{\pi}{30}\frac{dn}{dt} = M_t \tag{4.15}$$

In the preceding equations J, n, M_t, M_g, e_g, p_r, n_r, n_c, f_g, and f_c are the values of inertia, rotational speed, turbine mechanical moment, generator resistance moment, coefficient of load damping, rated power output, rated rotational speed, given rotational speed, generator frequency, and given frequency, respectively.

4.4 Perspective model

PHS technology is mature, offers the greatest roundtrip efficiency among all MES systems, is very fast in responding to load changes and operation mode alteration, and is not restricted in MW capacity. Also, the production cost of PHS systems is quite low due to the nature of how a PHS unit operates. However, the capital cost of PHS systems is significantly high. Unfortunately, there is not a promising solution for reducing the cost of PHS systems because, as previously mentioned, this technology is quite mature and has been under applied development for decades. This along with the environmental concerns associated with PHS systems, the fairly long time required for constructing a PHS plant, and technical challenges such as sediments are the drawbacks of this technology.

Considering all its positive and negative points, PHS technology will continue to be a key component of energy systems around the world. This is especially important now after the sharp growth of renewable energy systems worldwide and the planned roadmap towards pure renewable energy systems. However, the undeniable fact is that there is still a gap between the point that research and development activities have taken this technology to and what the market needs in an optimal situation. That is why studies on this concept and efforts to address its remaining challenges are continuing. PHS plants with underground reservoirs, undersea PHS systems, and so on are all examples of recent innovations in this area. The most critical needs in this area for future studies and developments could be highlighted as follows:

- Advances in designing the turbine and pump are required to further improve the cost-effectiveness and roundtrip efficiency of the system.
- It seems that digitalizing current energy storage technologies is required for different aspects of development. It is expected that development and advances in automation and information technology will shape the future of PHS plants.
- Environmental impact is one of the main concerns and limitations of the development of PHS plants mainly due to finding or creating

water reservoirs with height difference. Technical developments in environmental sciences will directly affect the methodologies by which the ecological impacts of existing and new PHS plants are assessed.

• Optimization of not only the processes of the system but also the trading and operation strategies of PHS plants in electricity markets when integrated with renewable power plants (such as wind turbines and solar power plants) could be effective in obtaining a better economic performance from such energy storage systems.

References

[1] A. Arabkoohsar, Combination of air-based high-temperature heat and power storage system with an organic Rankine cycle for an improved electricity efficiency. Appl. Therm. Eng. 167 (2020) 114762, https://doi.org/10.1016/j.applthermaleng.2019.114762.

[2] A. Arabkoohsar, G.B. Andresen, Dynamic energy, exergy and market modeling of a high temperature heat and power storage system. Energy 126 (2017), https://doi.org/10.1016/j.energy.2017.03.065.

[3] F. Geth, T. Brijs, J. Kathan, J. Driesen, R. Belmans, An overview of large-scale stationary electricity storage plants in Europe: current status and new developments. Renew. Sust. Energ. Rev. 52 (2015) 1212–1227, https://doi.org/10.1016/j.rser.2015.07.145.

[4] A. Arabkoohsar, M. Dremark-Larsen, R. Lorentzen, G.B. Andresen, Subcooled compressed air energy storage system for coproduction of heat, cooling and electricity. Appl. Energy 205 (2017) 602–614, https://doi.org/10.1016/J.APENERGY.2017.08.006.

[5] U.S. Department of Energy DOE OE Global Energy Storage Database, 2017.

[6] C.J. Yang, R.B. Jackson, Opportunities and barriers to pumped-hydro energy storage in the United States. Renew. Sust. Energ. Rev. 15 (2011) 839–844, https://doi.org/10.1016/j.rser.2010.09.020.

[7] Atle Harby, Julian Sauterleute, Magnus Korpås, Ånund Killingtveit, Eivind Solvang, Torbjørn Nielsen, Pumped Storage Hydropower, n.d. https://doi.org/10.1002/9783527673872.ch29.

[8] S. Koohi-Fayegh, M.A. Rosen, A review of energy storage types, applications and recent developments. J. Energy Storage 27 (2020), https://doi.org/10.1016/j.est.2019.101047.

[9] J. Menéndez, J.M. Fernández-Oro, M. Galdo, J. Loredo, Pumped-storage hydropower plants with underground reservoir: influence of air pressure on the efficiency of the Francis turbine and energy production. Renew. Energy 143 (2019) 1427–1438, https://doi.org/10.1016/j.renene.2019.05.099.

[10] D. Agar, M. Rasi, On the use of a laboratory-scale Pelton wheel water turbine in renewable energy education. Renew. Energy 33 (2008) 1517–1522, https://doi.org/10.1016/j.renene.2007.09.003.

[11] D.S. Benzon, G.A. Aggidis, J.S. Anagnostopoulos, Development of the Turgo impulse turbine: past and present. Appl. Energy 166 (2016) 1–18, https://doi.org/10.1016/j.apenergy.2015.12.091.

[12] K. Gaiser, P. Erickson, P. Stroeve, J.P. Delplanque, An experimental investigation of design parameters for pico-hydro Turgo turbines using a response surface methodology. Renew. Energy 85 (2016) 406–418, https://doi.org/10.1016/j.renene.2015.06.049.

[13] B.R. Cobb, K.V. Sharp, Impulse (Turgo and Pelton) turbine performance character-
     istics and their impact on pico-hydro installations. Renew. Energy 50 (2013)
     959–964, https://doi.org/10.1016/j.renene.2012.08.010.
[14] E. Ferrer, S. Le Clainche, Simple Models for Cross Flow Turbines. Springer Tracts in
     Mechanical EngineeringSpringer International Publishing, 2019, pp. 1–10, https://doi.
     org/10.1007/978-3-030-11887-7_1.
[15] E. Ferrer, S. Le Clainche, Flow Scales in Cross-Flow Turbines. Springer Tracts in
     Mechanical EngineeringSpringer International Publishing, 2015, https://doi.org/
     10.1007/978-3-319-16202-7_1.
[16] M. Ebrahimi, S. Duncan, M.R. Belmont, P. Kripakaran, G. Tabor, I. Moon,
     S. Djordjević, Flume experiments on the impact of a cross-flow turbine on an erodible
     bed. Renew. Energy 153 (2020) 1219–1225, https://doi.org/10.1016/j.renene.
     2020.02.073.
[17] M. Nazari-Heris, B. Mohammadi-Ivatloo, Design of small hydro generation systems.
     in: Distributed Generation Systems. Design, Operation and Grid Integration, Elsevier,
     2017, pp. 301–332, https://doi.org/10.1016/B978-0-12-804208-3.00006-6.
[18] S.C. Bhatia, Hydroelectric power. in: Advanced Renewable Energy Systems, Elsevier,
     2014, pp. 240–269, https://doi.org/10.1016/b978-1-78242-269-3.50010-3.
[19] M.P. Boyce, Advanced industrial gas turbines for power generation. in: Combined
     Cycle Systems for Near-Zero Emission Power Generation, Elsevier, 2012,
     pp. 44–102, https://doi.org/10.1533/9780857096180.44.
[20] P. Breeze, Chapter 8. Hydropower. in: Power Generation Technologies, third ed.,
     2019, pp. 173–201, https://doi.org/10.1016/B978-0-08-102631-1.00008-0.
[21] P. Breeze, Hydropower turbines. in: Hydropower, Elsevier, 2018, pp. 35–46, https://
     doi.org/10.1016/b978-0-12-812906-7.00004-1.
[22] E. Kadaj, R. Bosleman, Energy recovery devices in membrane desalination processes.
     in: Renewable Energy Powered Desalination Handbook. Application and Thermody-
     namics, Elsevier Inc., 2018, pp. 415–444, https://doi.org/10.1016/B978-0-12-
     815244-7.00011-8
[23] I. Samora, V. Hasmatuchi, C. Münch-Alligné, M.J. Franca, A.J. Schleiss, H.M.
     Ramos, Experimental characterization of a five blade tubular propeller turbine for pipe
     inline installation. Renew. Energy 95 (2016) 356–366, https://doi.org/10.1016/j.
     renene.2016.04.023.
[24] A.H. Elbatran, O.B. Yaakob, Y.M. Ahmed, H.M. Shabara, Operation, performance
     and economic analysis of low head micro-hydropower turbines for rural and remote
     areas: a review. Renew. Sust. Energ. Rev. 43 (2015) 40–50, https://doi.org/
     10.1016/j.rser.2014.11.045.
[25] G.A. Caxaria, D. de Mesquitae Sousa, H.M. Ramos, Small scale hydropower: generator
     analysis and optimization for water supply systems. in: Proceedings of the World
     Renewable Energy Congress, Sweden, 8–13 May, 2011, Linköping, Sweden, Lin-
     köping University Electronic Press, 2011, pp. 1386–1393, https://doi.org/10.3384/
     ecp110571386.
[26] D.K. Okot, Review of small hydropower technology. Renew. Sust. Energ. Rev.
     26 (2013) 515–520, https://doi.org/10.1016/j.rser.2013.05.006.
[27] M.S. Ramsey, Pumps. in: Practical Wellbore Hydraulics and Hole Cleaning, Elsevier,
     2019, pp. 271–282, https://doi.org/10.1016/B978-0-12-817088-5.00008-3.
[28] N. Sivakumar, D. Das, N.P. Padhy, Variable speed operation of reversible pump-
     turbines at Kadamparai pumped storage plant—a case study. Energy Convers. Manag.
     78 (2014) 96–104, https://doi.org/10.1016/j.enconman.2013.10.048.
[29] M. Ferrini, W. Borreani, G. Lomonaco, F. Magugliani, Design by theoretical and CFD
     analyses of a multi-blade screw pump evolving liquid lead for a generation IV LFR. Nucl.
     Eng. Des. 297 (2016) 276–290, https://doi.org/10.1016/j.nucengdes.2015.12.006.

[30] E. Corà, J. Jacques Fry, M. Bachhiesl, A. Schleiss, Hydropower Technologies: The State-of-the-Art, Hydropower Europe, 2019.

[31] A. Berrada, K. Loudiyi, I. Zorkani, Dynamic modeling and design considerations for gravity energy storage. J. Clean. Prod. 159 (2017) 336–345, https://doi.org/10.1016/j.jclepro.2017.05.054.

[32] T. Morstyn, M. Chilcott, M.D. McCulloch, Gravity energy storage with suspended weights for abandoned mine shafts. Appl. Energy 239 (2019) 201–206, https://doi.org/10.1016/j.apenergy.2019.01.226.

[33] J.D. Hunt, B. Zakeri, G. Falchetta, A. Nascimento, Y. Wada, K. Riahi, Mountain gravity energy storage: a new solution for closing the gap between existing short- and long-term storage technologies. Energy 190 (2020) 116419, https://doi.org/10.1016/j.energy.2019.116419.

[34] A. Berrada, K. Loudiyi, R. Garde, Dynamic modeling of gravity energy storage coupled with a PV energy plant. Energy 134 (2017) 323–335, https://doi.org/10.1016/j.energy.2017.06.029.

[35] G. Notton, D. Mistrushi, L. Stoyanov, P. Berberi, Operation of a photovoltaic-wind plant with a hydro pumping-storage for electricity peak-shaving in an island context. Sol. Energy 157 (2017) 20–34, https://doi.org/10.1016/j.solener.2017.08.016.

[36] X. Xu, W. Hu, D. Cao, Q. Huang, C. Chen, Z. Chen, Optimized sizing of a standalone PV-wind-hydropower station with pumped-storage installation hybrid energy system. Renew. Energy 147 (2020) 1418–1431, https://doi.org/10.1016/j.renene.2019.09.099.

[37] J.S. Anagnostopoulos, D.E. Papantonis, Simulation and size optimization of a pumped-storage power plant for the recovery of wind-farms rejected energy. Renew. Energy 33 (2008) 1685–1694, https://doi.org/10.1016/j.renene.2007.08.001.

[38] K. Weman, Power sources for arc welding. in: Welding Processes Handbook, second ed., Woodhead Publishing, 2012, pp. 51–62, https://doi.org/10.1533/9780857095183.51.

[39] Y. Yu, Y. Wang, F. Sun, The latest development of the motor/generator for the flywheel energy storage system. in: Proceddings of 2011 International Conference on Mechatronic Science, Electric Engineering and Computer (MEC), 2011, pp. 1228–1232, https://doi.org/10.1109/MEC.2011.6025689.

[40] British Electricity International, Station design and layout. in: Station Planning and Design, Elsevier, 1991, pp. 59–177, https://doi.org/10.1016/b978-0-08-040511-7.50009-7.

[41] G. Takacs, Use of ESP equipment in special conditions. in: Electrical Submersible Pumps Manual, Elsevier, 2009, pp. 119–186, https://doi.org/10.1016/b978-1-85617-557-9.00004-x.

[42] R. Montgomery, R. McDowall, Basics of electricity. in: Fundamentals of HVAC Control Systems, Elsevier, 2008, pp. 30–60, https://doi.org/10.1016/b978-0-08-055233-0.00002-9.

[43] S. Rehman, L.M. Al-Hadhrami, M.M. Alam, Pumped hydro energy storage system: a technological review. Renew. Sust. Energ. Rev. 44 (2015) 586–598, https://doi.org/10.1016/j.rser.2014.12.040.

[44] I. Kougias, S. Szabó, Pumped hydroelectric storage utilization assessment: forerunner of renewable energy integration or Trojan horse?. Energy 140 (2017) 318–329, https://doi.org/10.1016/j.energy.2017.08.106.

[45] P. Sullivan, W. Short, N. Blair, Modeling the benefits of storage technologies to wind power. Wind Eng. 32 (2008) 603–615, https://doi.org/10.1260/030952408787548820.

[46] B.C. Ummels, E. Pelgrum, W.L. Kling, Integration of large-scale wind power and use of energy storage in the Netherlands' electricity supply. in: IET Renewable Power Generation, 2008, pp. 34–46, https://doi.org/10.1049/iet-rpg:20070056.

[47] P. Denholm, M. Hand, Grid flexibility and storage required to achieve very high penetration of variable renewable electricity. Energy Policy 39 (2011) 1817–1830, https://doi.org/10.1016/j.enpol.2011.01.019.

[48] Up2Europe Hyperbole: European Project, 2013.

[49] A. Botterud, T. Levin, V. Koritarov, Pumped Storage Hydropower: Benefits for Grid Reliability and Integration of Variable Renewable Energy, Argonne National Laboratory, 2014.

[50] I. Kougias, G. Aggidis, F. Avellan, S. Deniz, U. Lundin, A. Moro, S. Muntean, D. Novara, J.I. Pérez-Díaz, E. Quaranta, P. Schild, N. Theodossiou, Analysis of emerging technologies in the hydropower sector. Renew. Sust. Energ. Rev. 113 (2019) 109257, https://doi.org/10.1016/j.rser.2019.109257.

[51] I. Iliev, C. Trivedi, O.G. Dahlhaug, Variable-speed operation of Francis turbines: a review of the perspectives and challenges. Renew. Sust. Energ. Rev. 103 (2019) 109–121, https://doi.org/10.1016/j.rser.2018.12.033.

[52] W. Yang, J. Yang, Advantage of variable-speed pumped storage plants for mitigating wind power variations: integrated modelling and performance assessment. Appl. Energy 237 (2019) 720–732, https://doi.org/10.1016/j.apenergy.2018.12.090.

[53] A. Vargas-Serrano, A. Hamann, S. Hedtke, C.M. Franck, G. Hug, Economic benefit analysis of retrofitting a fixed-speed pumped storage hydropower plant with an adjustable-speed machine. in: 2017 IEEE Manchester PowerTech, Powertech 2017, Institute of Electrical and Electronics Engineers Inc., 2017, https://doi.org/10.1109/PTC.2017.7981008

[54] P. Haney, T. Grider, E. Muljadi, H. Obermeyer, R. Robichaud, L. George, Cost estimation of a permanent magnet synchronous machine for use in adjustable speed-pumped storage hydropower. in: IEEE Power & Energy Society General Meeting, IEEE Computer Society, 2019, https://doi.org/10.1109/PESGM40551.2019.8973963.

[55] A.A. Salimi, A. Karimi, Y. Noorizadeh, Simultaneous operation of wind and pumped storage hydropower plants in a linearized security-constrained unit commitment model for high wind energy penetration. J. Energy Storage 22 (2019) 318–330, https://doi.org/10.1016/j.est.2019.02.026.

[56] J. Menéndez, J.M. Fernández-Oro, M. Galdo, J. Loredo, Efficiency analysis of underground pumped storage hydropower plants. J. Energy Storage 28 (2020) 101234, https://doi.org/10.1016/j.est.2020.101234.

[57] E. Pujades, P. Orban, S. Bodeux, P. Archambeau, S. Erpicum, A. Dassargues, Underground pumped storage hydropower plants using open pit mines: how do groundwater exchanges influence the efficiency?. Appl. Energy 190 (2017) 135–146, https://doi.org/10.1016/j.apenergy.2016.12.093.

[58] Z. Zhao, J. Yang, W. Yang, J. Hu, M. Chen, A coordinated optimization framework for flexible operation of pumped storage hydropower system: nonlinear modeling, strategy optimization and decision making. Energy Convers. Manag. 194 (2019) 75–93, https://doi.org/10.1016/j.enconman.2019.04.068.

[59] M.W. Tian, S.R. Yan, X.X. Tian, S. Nojavan, K. Jermsittiparsert, Risk and profit-based bidding and offering strategies for pumped hydro storage in the energy market. J. Clean. Prod. 256 (2020) 120715, https://doi.org/10.1016/j.jclepro.2020.120715.

[60] M. Ak, E. Kentel, S. Savasaneril, Quantifying the revenue gain of operating a cascade hydropower plant system as a pumped-storage hydropower system. Renew. Energy 139 (2019) 739–752, https://doi.org/10.1016/j.renene.2019.02.118.

[61] C. Cheng, C. Su, P. Wang, J. Shen, J. Lu, X. Wu, An MILP-based model for short-term peak shaving operation of pumped-storage hydropower plants serving multiple power grids. Energy 163 (2018) 722–733, https://doi.org/10.1016/j.energy.2018.08.077.

[62] D. Connolly, H. Lund, P. Finn, B.V. Mathiesen, M. Leahy, Practical operation strategies for pumped hydroelectric energy storage (PHES) utilising electricity price

arbitrage. Energy Policy 39 (2011) 4189–4196, https://doi.org/10.1016/j.enpol.2011.04.032.

[63] J. Jurasz, J. Mikulik, M. Krzywda, B. Ciapała, M. Janowski, Integrating a wind- and solar-powered hybrid to the power system by coupling it with a hydroelectric power station with pumping installation. Energy 144 (2018) 549–563, https://doi.org/10.1016/j.energy.2017.12.011.

[64] M. Daneshvar, B. Mohammadi-Ivatloo, K. Zare, S. Asadi, Two-stage stochastic programming model for optimal scheduling of the wind-thermal-hydropower-pumped storage system considering the flexibility assessment. Energy 193 (2020) 116657, https://doi.org/10.1016/j.energy.2019.116657.

[65] B. Xu, D. Chen, M. Venkateshkumar, Y. Xiao, Y. Yue, Y. Xing, P. Li, Modeling a pumped storage hydropower integrated to a hybrid power system with solar-wind power and its stability analysis. Appl. Energy 248 (2019) 446–462, https://doi.org/10.1016/j.apenergy.2019.04.125.

[66] C. Su, C. Cheng, P. Wang, J. Shen, X. Wu, Optimization model for long-distance integrated transmission of wind farms and pumped-storage hydropower plants. Appl. Energy 242 (2019) 285–293, https://doi.org/10.1016/j.apenergy.2019.03.080.

[67] X. Wang, L. Chen, Q. Chen, Y. Mei, H. Wang, Model and analysis of integrating wind and PV power in remote and core areas with small hydropower and pumped hydropower storage. Energies 11 (2018) 3459, https://doi.org/10.3390/en11123459.

[68] J. Dujardin, A. Kahl, B. Kruyt, S. Bartlett, M. Lehning, Interplay between photovoltaic, wind energy and storage hydropower in a fully renewable Switzerland. Energy 135 (2017) 513–525, https://doi.org/10.1016/j.energy.2017.06.092.

[69] J.I. Pérez-Díaz, J. Jiménez, Contribution of a pumped-storage hydropower plant to reduce the scheduling costs of an isolated power system with high wind power penetration. Energy 109 (2016) 92–104, https://doi.org/10.1016/j.energy.2016.04.014.

[70] P.D. Brown, J.A. Peças Lopes, M.A. Matos, Optimization of pumped storage capacity in an isolated power system with large renewable penetration. IEEE Trans. Power Syst. 23 (2008) 523–531, https://doi.org/10.1109/TPWRS.2008.919419.

[71] M.W. Murage, C.L. Anderson, Contribution of pumped hydro storage to integration of wind power in Kenya: an optimal control approach. Renew. Energy 63 (2014) 698–707, https://doi.org/10.1016/j.renene.2013.10.026.

[72] M. Kapsali, J.K. Kaldellis, Combining hydro and variable wind power generation by means of pumped-storage under economically viable terms. Appl. Energy 87 (2010) 3475–3485, https://doi.org/10.1016/j.apenergy.2010.05.026.

[73] D. Al Katsaprakakis, D.G. Christakis, K. Pavlopoylos, S. Stamataki, I. Dimitrelou, I. Stefanakis, P. Spanos, Introduction of a wind powered pumped storage system in the isolated insular power system of Karpathos-Kasos. Appl. Energy 97 (2012) 38–48, https://doi.org/10.1016/j.apenergy.2011.11.069.

[74] T. Ma, H. Yang, L. Lu, J. Peng, Pumped storage-based standalone photovoltaic power generation system: modeling and techno-economic optimization. Appl. Energy 137 (2015) 649–659, https://doi.org/10.1016/j.apenergy.2014.06.005.

[75] P. Javanbakht, S. Mohagheghi, M.G. Simoes, Transient performance analysis of a small-scale PV-PHS power plant fed by a SVPWM drive applied for a distribution system. in: 2013 IEEE Energy Conversion Congress and Exposition. ECCE 2013, 2013, pp. 4532–4539, https://doi.org/10.1109/ECCE.2013.6647307.

[76] J. Margeta, Z. Glasnovic, Theoretical settings of photovoltaic-hydro energy system for sustainable energy production. Sol. Energy 86 (2012) 972–982, https://doi.org/10.1016/j.solener.2012.01.007.

[77] D. Manolakos, G. Papadakis, D. Papantonis, S. Kyritsis, A stand-alone photovoltaic power system for remote villages using pumped water energy storage. Energy 29 (2004) 57–69, https://doi.org/10.1016/j.energy.2003.08.008.

[78] M. Melikoglu, Pumped hydroelectric energy storage: analysing global development and assessing potential applications in Turkey based on vision 2023 hydroelectricity wind and solar energy targets. Renew. Sust. Energ. Rev. 72 (2017) 146–153, https://doi.org/10.1016/j.rser.2017.01.060.

[79] F. Winde, F. Kaiser, E. Erasmus, Exploring the use of deep level gold mines in South Africa for underground pumped hydroelectric energy storage schemes. Renew. Sust. Energ. Rev. 78 (2017) 668–682, https://doi.org/10.1016/j.rser.2017.04.116.

[80] A. Karimi, S.L. Heydari, F. Kouchakmohseni, M. Naghiloo, Scheduling and value of pumped storage hydropower plant in Iran power grid based on fuel-saving in thermal units. J. Energy Storage 24 (2019) 100753, https://doi.org/10.1016/j.est.2019.04.027.

[81] Z. Dong, J. Tan, E. Muljadi, R. Nelms, M. Jacobson, Impacts of ternary-pumped storage hydropower on U.S. western interconenction with extremely high renewable penetrations. in: IEEE Power and Energy Society General Meeting, IEEE Computer Society, 2019, https://doi.org/10.1109/PESGM40551.2019.8973787.

[82] X. Lu, S. Wang, A GIS-based assessment of Tibet's potential for pumped hydropower energy storage. Renew. Sust. Energ. Rev. 69 (2017) 1045–1054, https://doi.org/10.1016/j.rser.2016.09.089.

[83] M. Gimeno-Gutiérrez, R. Lacal-Arántegui, Assessment of the European potential for pumped hydropower energy storage based on two existing reservoirs. Renew. Energy 75 (2015) 856–868, https://doi.org/10.1016/j.renene.2014.10.068.

[84] K. Kusakana, Optimal operation control of pumped hydro storage in the south african electricity market. in: Energy Procedia, Elsevier Ltd, 2017, pp. 804–810, https://doi.org/10.1016/j.egypro.2017.12.766.

[85] Y. Ko, G. Choi, S. Lee, S. Kim, Economic analysis of pumped hydro storage under Korean governmental expansion plan for renewable energy. in: Energy Reports, Elsevier Ltd, 2020, pp. 214–220, https://doi.org/10.1016/j.egyr.2019.08.047.

[86] J.K. Kaldellis, M. Kapsali, K.A. Kavadias, Energy balance analysis of wind-based pumped hydro storage systems in remote island electrical networks. Appl. Energy 87 (2010) 2427–2437, https://doi.org/10.1016/j.apenergy.2010.02.016.

[87] M. Kapsali, J.S. Anagnostopoulos, J.K. Kaldellis, Wind powered pumped-hydro storage systems for remote islands: a complete sensitivity analysis based on economic perspectives. Appl. Energy 99 (2012) 430–444, https://doi.org/10.1016/j.apenergy.2012.05.054.

[88] J.S. Anagnostopoulos, D.E. Papantonis, Study of pumped storage schemes to support high RES penetration in the electric power system of Greece. Energy 45 (2012) 416–423, https://doi.org/10.1016/j.energy.2012.02.031.

[89] A.K. Varkani, A. Daraeepour, H. Monsef, A new self-scheduling strategy for integrated operation of wind and pumped-storage power plants in power markets. Appl. Energy 88 (2011) 5002–5012, https://doi.org/10.1016/j.apenergy.2011.06.043.

[90] E.B. Wylie, V.L. Streeter, Fluid Transients, McGraw Hill, New York, NY, 1978.

[91] N. Mousavi, G. Kothapalli, D. Habibi, M. Khiadani, C.K. Das, An improved mathematical model for a pumped hydro storage system considering electrical, mechanical, and hydraulic losses. Appl. Energy 247 (2019) 228–236, https://doi.org/10.1016/j.apenergy.2019.03.015.

[92] W. Yang, J. Yang, W. Guo, W. Zeng, C. Wang, L. Saarinen, P. Norrlund, A mathematical model and its application for hydro power units under different operating conditions. Energies 8 (2015) 10260–10275, https://doi.org/10.3390/en80910260.

CHAPTER FIVE

Flywheel energy storage

Ahmad Arabkoohsar and Meisam Sadi

Department of Energy Technology, Aalborg University, Esbjerg, Denmark

Abstract

A flywheel energy storage (FES) system is an electricity storage technology under the category of mechanical energy storage (MES) systems that is most appropriate for small- and medium-scale uses and shorter period applications. In an FES system, the surplus electricity is stored in a high rotational velocity disk-shaped flywheel. The stored energy in the form of kinetic energy will be later used to drive a generator and thereby produce electrical power. This chapter gives a fundamental understanding of the flywheel and FES system mechanism followed by a history of the development of this technology in the literature and practice over recent years. It also discusses FES system applications in several fields of science and technology, advantages and disadvantages, design concept, and components. Finally, it presents the mathematical formulation governing the system as well as the future perspectives of the technology.

5.1 Fundamentals

This section discusses general configuration/information, pros and cons, and the real-life applications of flywheel energy storage (FES) technology.

5.1.1 The FES system

As discussed in the previous chapters, energy storage systems are becoming increasingly popular, especially for public power utilities and a wide range of applications in power systems [1]. Among the existing energy storage solutions, a majority are appropriate for short-term storage and a few are appropriate for long-term use and large capacities. The latter class of storage solutions is mainly under the category of mechanical energy storage (MES) systems, most of which are in the development phase. However, this does not mean that all MES systems are well developed and suitable for long-term and large-scale uses [2]. FES technology is one of the maturest MES systems, known as a clean storage technology that is highly efficient and appropriate for short storage times and low and medium capacities [3]. The FES system works in a high vacuum situation and is identified by low friction resistance, low wind resistance, low maintenance, and has a great lifecycle and no effects on the surrounding environment [4].

Like any other energy storage system, an FES system receives the available extra power and stores it to give it back as electric power whenever required [5]. When an FES system is working it has two modes of charging (receiving power) and discharging (giving the stored power back) as well as a third standby mode. In the charging phase of an FES system, a shaft connected to the flywheel disk receives the available power to rotate at a very fast velocity. This means that the supplied electricity is transformed into inertial energy and stored in the system as rotational power. In the discharging process, the power is taken back from the flywheel to produce power and, as a result, the rotor's rotational velocity decreases simply based on the energy conservation law. In the standby mode, the system neither receives power from any source nor drives the generator for power generation.

For a better understanding of the energy storage mechanism in an FES system, the flywheel itself should be first introduced. A flywheel is a mass storage machine, a mechanical device like a disk with a definite amount of mass designed to effectively accumulate the rotational power in kinetic form. In other words, and as shown in Fig. 5.1, a flywheel is a massive spinning disk mounted on a shaft that speeds up when receiving power.

Flywheels endure variations in rotational velocity by their moment of inertia. Knowing this ability of flywheels in storing energy in the form of

Fig. 5.1 Schematic of a flywheel.

kinetic inertial energy, the FES system was then introduced. The beginning of the implementation of flywheels in electric equipment as an energy storage device goes back to the 1960s [6].

In addition to the flywheel (which is the heart of this storage system as the design of the machine is based on the mechanical inertia mechanism via storing kinetic energy in the form of a rotational mass), there are five other main components of an FES system. These are the power electronics, the motor-generator, the mechanical or magnetic bearing, the external inductor, and a vacuum pump.

The flywheel has to be connected to an electrical motor-generator, through which the system is able to charge and discharge. The motor-generator would be the same machine in which the input and output are reversed. The electrical motor-generator should be designed so that it has minimum size with greatest rotational velocity. Typical electrical machines used in FES systems include induction machines (IMs), switched reluctance machines (SRMs), and permanent magnet synchronous machines (PMSMs) [7]. The flywheel is coupled to one of these motor-generator machines using an AC/AC matrix converter [8, 9]. Table 5.1 presents a comparison of these machines as well as their advantages and disadvantages.

The power electronics part includes a three-phase insulated-gate bipolar transistor-based pulse-width modulation inventor-rectifier. An insulated-

Table 5.1 Comparison of different kinds of electrical machines [10].

	Advantages	Disadvantages
Permanent magnet synchronous machines	Free of actuation requirements Simple rotor design High performance index High energy density	Demagnetization possibility Need for support against centrifugal forces Sensitive to thermal stresses
Induction machines	Free of demagnetization Possible control on actuation field Low cost	Insufficient overload capability Extensive maintenance needs Complex rotor design External actuation needs
Switched reluctance machines	Much robust Free of idle losses	Low energy factor and energy density External actuation needs High rotor eddy current losses

gate bipolar transistor is a solid-state accessory with the capability to deal with voltages up to 6.7 kV, electrical currents up to 1.2 kA, and large switching frequencies.

The mechanical or magnetic bearings of the FES system are located in a vacuum space. For angular velocity less than 20,000 rpm, mechanical bearings are the preferred type, whereas in the case of angular velocity greater than 40,000 rpm, magnetic bearings would be the correct selection [11]. Mechanical bearings require lower initial costs but have relatively high friction losses. For high angular velocity (the enclosure pressure is low and therefore lubrication is difficult), mechanical bearings are unacceptable because of the lubrication issue. The evacuation reduces friction and increases efficiency.

As high-velocity motors provide small inductances with a small number of stator turns, using external inductances in a series configuration is essential to cut down the high total harmonic distortion of the system in the charging process.

Finally, as mentioned before, the FES system components are in a vacuum space and thus a vacuum pump is required to provide the required vacuity of the system.

Fig. 5.2 presents a schematic of an FES system that includes all these components as well as some other auxiliary parts.

The surplus power to be stored (probably from a renewable source of energy) might be sent to the DC motor to rotate the shaft. The type of motor might be a DC brushless, ironless armature-type motor. The maximum power point tracker would be used to manage and increase the generated energy and improve the cost-effectiveness. For example, in solar photovoltaic panels, fluctuating produced electrical power is typically inconsistent with the specification of the FES system and load, inducing an ineffective procedure. The generator should be designed to respond to the huge demands of different loads. The vacuum enclosure secures a comparatively high storage time before

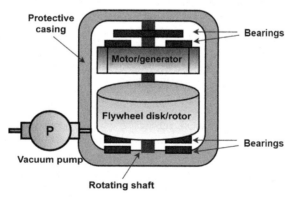

Fig. 5.2 Illustration of an FES system.

internal losses dissipate the power of the flywheel. For security considerations, the package can be located underground. The system should be operated with the frequency controlled at 60 Hz. Under specified conditions, provisions can be made for the cold system startup from the solar PV input power. By using DC input from the external source, DC control power drives until the flywheel reaches the rotational velocity. To avoid decreasing the input voltage, the maximum power tracking feature of the input power circuit is used.

The circuits of the motor-generator electronics are shown in Fig. 5.3 and Fig. 5.4. As shown in Fig. 5.3, a full-bridge DC-AC inverter is connected, stimulating the flywheel motor and charging the system. To generate power in the discharging process, the flywheel generator through an AC-AC converter circuit, as shown in Fig. 5.4, is connected to the grid.

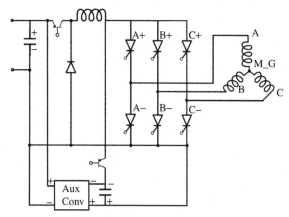

Fig. 5.3 The FES system motor power electronics.

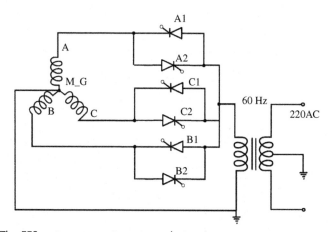

Fig. 5.4 The FES system generator power electronics.

5.1.2 Pros and cons

To better understand the performance of the FES system and to correctly assess the applicability of the technology, it is necessary to review the related advantages and disadvantages. The energy efficiency of the FES system is claimed to be greater than 85% [12]. For this system, the maintenance costs are low and the lifetime is fairly long. No toxic or greenhouse gases or dangerous materials are produced during operation, hence it is environmentally friendly. The storage capacity is independent of the temperature variations. The number of charging-discharging cycles and depth of discharge do not affect the life cycle of the flywheel. By using magnetic levitation, the lifetime of the FES system increases. The FES system also has a fast response time, which makes it capable of agile reaction to the fast ramps of renewable sources. In special types of FES systems, like those with permanent magnet motor-generators, the efficiency is greater and the size of the system is smaller than other FES systems. This type of machine also has less friction losses and less winding inductances, making it more useful for applications with a fast response requirement. Also, bearings have even less losses due to magnetic levitation and thus do not wear down, making their lifetime longer.

Despite the advantages of the FES system, there are some disadvantages as well. This system is a low-capacity storage technology with large power loss and low energy density. Due to the high velocity of the rotor and its fracture risk, the expenses of the system's security are high. Size and tolerance considerations at high angular velocities are big concerns about the system. There are also some limitations for the materials to be used in the manufacturing of the flywheel. Materials have specified mechanical stresses and fatigue limits and not many of them can endure such high angular velocities. The FES system is slow in discharging the stored power and has a relatively low power rating. Being slow in discharging the stored energy, this technology cannot support an energy market for several hours. FES technology is not as mature as chemical battery technology and therefore cannot compete with the other currently available storage technologies in the market.

Outlining the pros and cons of FES technology, one could compare this system with other electricity storage systems to better understand its strengths and weaknesses. Table 5.2 compares a few important features of the FES system with other energy storage systems. The FES system (although being an MES technology), together with supercapacitors and superconduction magnetic energy storage systems, has the lowest response

Table 5.2 Comparison of energy storage methods in terms of energy capacity, power rating, and response time [13].

Technology		Nominal capacity	Discharge time	Response time	Efficiency	Lifetime (years)
Flywheels		10 kW–20 MW	1 s–1 h	5–10 ms	85%–95%	20
Pumped hydroelectric energy storage		100–4000 MW	6–24 h	10 s–3 min	65%–85%	30–75
Compressed air energy storage		25–30,000 MW	4–24 h	3–15 min	50%–85%	20–40
Battery	Flow batteries	25 kW–10 MW	1–8 h		65%–85%	
	Lithium-ion batteries	10 kW–10 MW	10 min–1 h	30–100 ms	85%–90%	2–10
	Lead-acid	50 kW–30 MW	15 min–4 h		70%–80%	
	Sodium-sulfur	50 kW–30 MW	1–8 h		75%–90%	
Supercapacitors		10 kW–1 MW	1 s–1 min	5–10 ms	85%–95%	40
Superconduction magnetic energy storage		1 MW–100 MW	1 s–1 min	5–10 ms	85%–95%	30–40

time in the range of 5–10 ms. Its efficiency is also comparable to batteries and way greater than other MES systems. The useful lifetime of the FES system is reported to be about 20 years, which is way longer than batteries but shorter than that of compressed air energy storage and pumped hydroelectric energy storage. On the other hand, the discharge time of the FES system is in the range of 1 s to 1 h, which is similar to that of batteries, but not comparable with the other MES systems. The minimum typical nominal power is also associated with the FES system together with lithium-ion batteries and supercapacitors.

We should also mention that the utilization of electrochemical batteries in uninterruptable power supplies and power quality system applications, which need a large number of charging and discharging cycles, is highly uncomfortable and negatively affected by unsatisfactory life cycle [14]. For example, for power quality systems, there are a lot of disturbances that should be responded to in just a few seconds. Hence, due to the longer life cycle and instantaneous response time, FES systems can efficiently handle such disturbances and possess excellence over batteries [15, 16]. For making a quantitative comparison, it should be noted that for a battery system with only one charge and discharge cycle per day, it is unlikely to be able to operate for 10 years (i.e., 3650 cycles). Of course, if the depth of discharge is considered low and the temperature conditions are arranged in the favorable range, batteries would be able to perform their function in the specified period; however, it is important to note that to compensate for the low depth of discharge, the battery storage capacity should be two to five times larger than the designed storage capacity. The result of such an increased storage capacity is clearly a greater initial investment cost.

5.1.3 Applications

The FES system is a popular and accepted technology in large numbers of applications, but mainly for relatively smaller-scale uses. It is commonly used to balance the demand, production, and frequency in electricity networks. In this case, the FES unit is considered a potential resource for fast power injection in load frequency control [17]. To perform this, battery energy storage can also be added to the FES system to make a hybrid energy storage system [18]. When the range of storage is from a few kW to several tens of MW, the FES system (in multiple units for larger scale) will be beneficial. For this type of application, the FES-based system is comparable with battery storage plants [19]. In the following, we review some real-life applications of FES systems.

Beacon Power completed the world's largest FES system at a cost of $60 m in Stephentown, New York. This 20-MW FES system marks a milestone in this area [20]. This facility implements 200 carbon fiber flywheels levitated in a vacuum enclosure and spun at a maximum velocity of 16,000 rpm to accumulate extra power and improve balance of the supply to the electricity network. The flywheels get extra energy and can regularly support the grid by the power of 1 MW for 15 min. The strategy is taken instead of an auxiliary natural gas plant that would generate power to regulate the supply and demand in the network. In 2012, Rockland Capital obtained the assets of the Beacon Power Corporation and developed a second flywheel storage facility with a capacity of 20 MW in Pennsylvania [20]. NASA's efforts to find a reliable alternative for batteries in spacecrafts for space exploitation purposes resulted in launching the NASA G2 FES module. Fig. 5.5 presents a picture of this FES module.

An FES system and corresponding conversion utilities have been implemented at the Massachusetts Institute of Technology, Lincoln Laboratory, to serve as a connection between a PV field and an AC load [21]. Most FES systems use electricity to charge and recharge the system, while facilities that use straight mechanical energy as input are under consideration.

Fig. 5.5 NASA's G2 FES module. *From https://en.wikipedia.org/wiki/File:G2_front2.jpg.*

An FES system has also been integrated with a variable speed wind generator to overcome the intermittent nature of power generation [22, 23].

The transportation sector is also one of the largest consumers of FES systems. The FES system has been implemented in buses, trains, cars, and so on. In public transportation, power is regained from the vehicle, a truck or train, when the brake is utilized and this energy is captured by an FES unit. This energy is reused for accelerating the vehicle whenever required [24]. In July 2014, GKN acquired Williams Hybrid Power division with the intention of providing 500 electric FES systems to metropolitan bus operators [25]. In September 2014, the Oxford Bus Company introduced the hybrid power gyrodrive bus in which an electric FES absorbs energy as the car brakes and then uses it to operate an electric motor, increase the power, and decrease fuel consumption [26]. Rupp et al. [27] introduced an FES system in a light rail transit train and estimated that the maximum energy savings and costs to be 31% and 11%, respectively. In motorsports applications, the recovered energy is used to enhance acceleration instead of decreasing CO_2 emission, while the FES unit can be utilized to enhance the fuel efficiency of the vehicle, too [28].

The FES system is used in pulse power applications as well, where the system feels the necessity for a large amount of energy for a short time, from milliseconds to a few seconds. These applicatios include weapons that require a large amount of power, aircraft powertrains, and shipboard power systems.

5.2 State-of-the-art

The utilization of stored energy in the FES system is not a brand-new technology. In this section, we take a glance into the history and current status of FES system technologies.

In the eighteenth century, during the Industrial Revolution and after the appearance of steam engines, two important developments occurred in the industry, namely, the use of flywheels in steam engines and the widespread use of metals instead of woods in the construction of equipment. The industry benefited from this advancement and cast iron was thereafter used in the manufacturing of very large flywheels with curved spokes. Now, owing to the large mass of cast iron in a certain volume, a greater moment of inertia could be achieved. However, the rotational velocity of the flywheel was still low. Thus, with the advancement in material and the need for greater rotational velocity, cast iron was replaced with the other types of materials with greater strength [29]. Then, flywheels were broadly employed on ships, trains, and in several other industries [30]. Thereafter, several configurations of flywheels were used in different industries over time. However, after advancements in analyzing the rotational stresses of the rotors and flywheels and development in metallurgy science, significant signs of progress began to appear in the early twentieth century. It was at this time that the flywheel was known as a potential energy storage vehicle [10]. The first application of an FES system in transportation was the Gyrobus, supplied by a 1500-kg flywheel, manufactured in Switzerland in 1950 [31]. In the following decades, FES technology was more used in electrical automobiles, for energy back up in different situations and industries, and space operations [32].

In recent years, knowing the potential of the FES system and its pros and cons, there has been a certain focus on developing better FES systems via improving parts for specific objectives. Today, fiber composite is generally used in manufacturing the rotors in FES systems. Many researchers reviewed different mechanisms that have been utilized in making up fiber composites for flywheel shafts [33–36]. Due to the high specific energy of composites compared to metals, a metal flywheel is preferred to composite flywheels [37]. The failure of fiber composite rotors are affected by a few factors. In single-material, circumferentially wound composite rims, the thickness between the inner and the outer part of the composite is a limitation. Composite structures with radially thin rims are conveniently fabricated with cheap expenditure, though thicker rims cannot be fabricated as a consequence of their finite radial strength [38]. It has been shown that for a thicker rim, the radial strain-to-failure ratio should increase and the radial modulus of elasticity of a filament wound composite rim should be reduced. It is also demonstrated that employing multi-ring filament wound rims decreases radial stress. Moreover, to have a very large change in stress distribution, two rings of different materials can be used in the rotor [39]. The radial strength might be enhanced by using layered laminates, woven fabrics, and other materials [40, 41]. In summary, it could be noted that fiber composite utilization in the fabrication of flywheels has improved the performance of FES systems. To enhance the performance of the FES system, Saidi and Djebli [42] tested three different materials: high-strength carbon fiber, Kevlar, and high-performance glass. They concluded that the high-strength carbon fiber is the best material that resists greater stresses. For obtaining greater rotational velocity and greater energy density, plain profiling woven carbon-fiber-reinforced composites were also tested in a few pieces of research [43–45]. These types of composites have favorable biaxial strength along radial and circumferential directions.

Applied and research work has been done to determine the optimum design of flywheels. These efforts have been mainly based on different criteria and thus plenty of different approaches have been proposed. Kress [46] employed an overall numerical shape optimization method with a 2D finite element method to obtain the best thickness distribution along the radius of a centrally bored flywheel. Here the objective was to approach an even stress distribution. He also carried out an analytical solution based on Stodola's model for an evenly stressed turbine disk. Chern and Prager [47] focused on the minimum-weight design of rotating disks to obtain the best shape from the viewpoint of uniform strength. They implemented a new constraint of the specified value for radial displacement. They concluded that the developed disks cannot satisfy the condition of uniform strength, while the mass is only slightly less than those disks with uniform strength. By using locally available materials, Okou et al. [48] proposed novel shape profiles based on Berger, Porat, and Stodola's models. One of the manufactured flywheels was certified on a rig with an axial flux permanent magnet machine.

Owing to advancements in bearings for electrical machines, active and passive magnetic bearings for FES systems have received special attention in recent years [49]. The application of low-velocity magnetic bearings was presented in the 1980s [50]. Specifically, advancement in magnetic bearings and materials keep the FES system as a competitive solution among energy storage systems [51, 52]. In the application of permanent magnetic bearing, there are nonsynchronous low-frequency vibrations that need to be controlled and damped [53, 54]. Hiroshima et al. [55] studied the effect of joining the disk and a driveshaft on the serious vibration at high angular velocities of rotating flywheels. They indicated that even at a high angular velocity, a resin ring embedded between the disk and rotor could make a stable joining. Qiu and Jiang [56] indicated that a magnetic pendulum tuned mass damper can be effective for this problem in a high-capacity FES system. Xiang and Wong [57] introduced a detailed relationship between the vibration specifications of the magnetically suspended rotor and system parameters. A rotor levitated radially by active magnetic bearings and axially by a permanent magnetic bearing were experimentally tested in [49, 58]. They indicated that the simulation well predicted the dynamics of the rotating shaft. To decrease the disturbances in the electric network, Miyamoto et al. [59] proposed an approach to connect the electric network of a squirrel cage induction generator rotor to a cylindrical flywheel. Miyazaki et al. [60] developed the world's largest-class FES system with a superconducting magnetic bearing. This system is capable of supporting the flywheel with a mass of 4000 kg.

Due to these advances, energy efficiencies up to 90% and greater energy densities can now be expected from the new generation of FES systems [61]. With storage potential of up to 500 MJ and power domain of kW to GW, FES systems can accomplish many necessary energy storage functions [22].

Salahuddin et al. [62] correlated graphically energy storage technologies and reviewed the advantages and disadvantages of the FES systems. Farhadi et al. [63] studied comparatively the application of an FES system for high-power cases. Kenny et al. [64] investigated a new control method for the charge and discharge processes of an FES system for space applications. Both simulation analysis and experimental investigation were carried out and indicated the fruitful operation associated with an FES system up to the rated velocity of 60,000 rpm. Pena-Alzola et al. [65] summarized flywheel-based energy storage systems. They stated that the FES system is advisable for a large number of charge and discharge courses with medium to high power, which occurs in a few seconds. Yu et al. [66] discussed the fundamentals of FES systems, investigated the key techniques for the development of motor-generators of such systems, and outlined the advancements in the components of FES systems. Okou et al. [48] performed a whole-life costing investigation to analyze the energy costs between an FES system and a lead-acid battery energy storage unit. Results showed that the integration of the flywheel into solar systems decreases the expenditure up to 37% per kWh. Soomro et al. [7] compared synchronous and induction machines used in flywheel energy storage systems for microgrid applications. This is because having an efficient motor-generator machine is one of the most important points in the energy conversion of FES systems [67]. The advantages and disadvantages of different types of machines are discussed in a few articles [68, 69]. Makbul et al. [70] analyzed the impact of utilizing a flywheel on power generation, energy cost, and net present cost for certain configurations of a hybrid system. They concluded that a hybrid PV–diesel system with an FES unit is more economical than one without an FES system. There are also many works on the investigations of the application of FES technology in wind systems. In these articles, the application of FES systems is considered from different aspects including control systems, type of electrical machine, effect of magnetic bearing, and others [71–75].

5.3 Mathematical model

When the design of an FES system is planned, two steps should be performed to ensure everything is running well. First, it should be estimated how much energy should be provided to smooth the power network fluctuations in the favorable defined level. Second, the mass of the flywheel that stores/provides that required energy should be determined. In this section, we present the required mathematical formulations for making general modeling of an FES system. This includes the definition of the fundamental parameters such as geometry, stored energy, stresses, and tensile strengths.

5.3.1 Fundamental equations

The mass moment of inertia or the moment of inertia of a rigid body, which is defined concerning a specific rotation axis, is a load that actuates the torque required for a desired angular acceleration about a rotational axis; similar to how mass regulates the force required for a favorable acceleration. The present value of energy and its period is influenced by two parameters: the mass and the velocity squared. The following equation presents the amount of kinetic energy of the disk-shaped flywheel as a function of the rotational velocity and moment of inertia.

$$E_k = \frac{1}{2}I\omega^2 \tag{5.1}$$

E_k, I, and ω denote the kinetic energy, the moment of inertia, and rotational velocity, respectively. According to the equation, the kinetic energy is a function of the moment of inertia and square of angular velocity. Low-velocity flywheels utilize the mass that they are equipped with, whereas a higher mass reserves higher energy. Large-velocity flywheels utilize the velocity they are turning where the greater the velocity, the greater the energy. The moment of inertia of a point mass with respect to an axis is obtained by multiplying the mass of the body by the square of its distance from the axis. This schematic and the related formulations of the flywheel motion are shown in Fig. 5.6 and presented below:

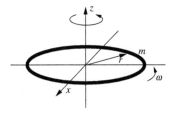

Fig. 5.6 Schematic of a rotating disk with an angular velocity of ω.

$$I = \int r^2 \, dm \qquad (5.2)$$

Where r is the length between the specified axis and the mass. In the International System of Units (SI), m is expressed in kilograms and distance in meters, and the moment of inertia has the dimension $\text{kg}\,\text{m}^2$. For a flywheel with the shape of a thin rim in which the mass is concentrated at the infinitely thin outer part of it, Eqs. (5.1) and (5.2) can be simplified as:

$$I = mr^2 \; \left[\text{kg}\,\text{m}^2\right] \qquad (5.3)$$

$$E_k = \frac{1}{2}mr^2\omega^2 \; [\text{J}] \qquad (5.4)$$

Then the question is "how much is the strength of the flywheel material?" The ultimate tensile strength makes limitation for the amount of stored energy in the flywheel. The stresses should not be greater than the ultimate tensile strength. The limiting stress in a slim rim is the tangential stress, which can be presented as [32]:

$$\sigma_{max} = \rho r^2 \omega^2 \; \left[\text{N}/\text{m}^2\right] \qquad (5.5)$$

The maximum tensile strength and the material density are denoted by σ_{max} and ρ. Then, the specific energy for an individual material, $e_{k,v}$ and the maximum energy density $e_{k,m}$ can be obtained.

$$e_{k,v} = \frac{1}{2}\sigma_{max} r^2 \omega^2 \; \left[\text{J}/\text{m}^3\right] \qquad (5.6)$$

$$e_{k,m} = \frac{1}{2}\frac{\sigma_{max}}{\rho} \; [\text{J}/\text{kg}] \qquad (5.7)$$

The equation for the specific energy for an individual material states that for having a high energy density, the flywheel material should have high tensile density. The maximum energy density states that a high strength material with low density is the best choice in manufacturing rotors. As explained before, the factor of 0.5 in Eqs. (5.6) and (5.7) is only valid for a rim-shaped flywheel. A more universal definition for $e_{k,v}$ and $e_{k,m}$ can be presented as:

$$e_{k,v} = K\sigma_{max} r^2 \omega^2 \; \left[\text{J}/\text{m}^3\right] \qquad (5.8)$$

$$e_{k,m} = K\frac{\sigma_{max}}{\rho} \; [\text{J}/\text{kg}] \qquad (5.9)$$

In which K denotes shape factor and means as a measure of how efficient the material of the flywheel is used. More information about the shape factor and the derivation method are gathered in [76]. The most typical shapes of a flywheel and the corresponding shape factor are shown in Fig. 5.7.

Generally, the material of flywheel plays a fundamental role in evaluating the specific energy value. Materials such as carbon fibers have higher specific energy and are the best selection for flywheels. These materials possess anisotropic properties. This means that in different directions, the characteristics of the material are variable. Table 5.3 presents a few general flywheel materials [76].

5.3.2 The formulation for optimum design of flywheel

Consider a constant-thickness flywheel (with thickness t and radius r) without bore at the center. The configuration of this flywheel is presented

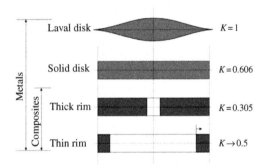

Fig. 5.7 Shape factor of typical flywheel shapes [77].

Table 5.3 Flywheel materials [78].

Material	Density (kg/m³)	Strength (MPa)	Theoretical specific energy (Wh/kg)
High-strength steel	7800	2500	90
Glass fiber	2600	2400	250
GFK (fiber in matrix)	2100	1500	200
Aramide fiber	1450	2800	540
AFK (fiber in matrix)	1400	1800	360
Carbon fiber	1800	7050	1090
CFK (fiber in matrix)	1500	4500	830

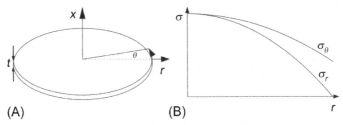

(A)   (B)

Fig. 5.8 Configuration of a constant-thickness flywheel (A), and the distribution of the normal and circumferential stresses (B) [46].

in Fig. 5.8A. The distribution of the normal stresses σ_r and circumferential stresses σ_θ are shown in Fig. 5.8B. The normal stress in radial direction σ_r must meet the natural boundary condition at the rim. The figure shows that the circumferential stresses are somewhat larger than the radial stress. Both normal and circumferential stresses decrease with increasing distance from the center. At the center of the rim, no difference is observed in the value of stresses.

For the central disk of the turbine, which is a link between blades and the central shaft, the design based on optimum strength is examined without considering a central hole.

For the configuration shown in Fig. 5.9, the equilibrium of forces in the radial direction can be expressed as [46]:

$$(\sigma_r t)'_r + \frac{t}{r}(\sigma_r - \sigma_\theta) + tr\omega^2\varrho = 0 \qquad (5.10)$$

both normal stresses σ_r and circumferential stresses σ_θ are considered constant [46]:

$$\sigma_r = \sigma_\theta = \sigma \qquad (5.11)$$

By considering the preceding equation, the general equilibrium equation can be presented as shown here. In this equation, the thickness is a dependent parameter [46].

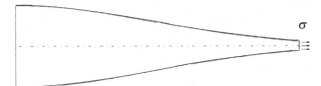

Fig. 5.9 Optimum turbine disk according to Stodola's design; the disk is without a central hole [46].

$$(t)'_r + \frac{tr\omega^2 \varrho}{\sigma} = 0 \tag{5.12}$$

By the solution of this equation, the most uniform shape for flywheel based on the stress distribution will be:

$$t = t_0 e^{-\frac{\omega^2 \varrho}{2\sigma} r^2} \tag{5.13}$$

It should be noted that the differential equation presented in Eq. (5.10) is true for the low values of reference thickness in comparison with the disk radius. This solution was achieved for a disk by considering no bore at its center, whereas flywheels have a central bore, which is used to connect them to the rotating shaft. To find the most even stress distribution for such a flywheel with a central bore, a finite volume method-based structural model can give the following equations in the axial and radial directions:

$$\tau_{rx'r} + \sigma_{x'x} + r^{-1}\tau_{rx} = 0 \tag{5.14}$$

$$\sigma_{r'r} + \tau_{rx'x} + r^{-1}(\sigma_r - \sigma_\theta) + r\omega^2 \varrho = 0 \tag{5.15}$$

By employing the principle of virtual displacements, we have:

$$\int_\Omega \left(\delta\omega'_r r\sigma_r + \delta\omega'_x r\tau_{rx} + \delta\omega\sigma_\theta + \delta u'_r r\tau_{rx} + \delta u'_x r\tau\sigma_x \right) dxdr$$

$$= \int_\Omega \delta\omega r^2 \omega^2 \varrho dxdr \tag{5.16}$$

By applying the constitutive equations for isotropic materials, the equation becomes:

$$\int_\Omega \delta\omega'_r r \left(C_{13}u'_x + C_{23}\frac{\omega}{r} + C_{33}\omega'_r \right) + \delta\omega'_x r \left(C_{55}\omega'_x + C_{55}u'_r \right)$$

$$+ \delta\omega_r \left(C_{12}u'_x + C_{22}\frac{\omega}{r} + C_{23}\omega'_r \right) + \delta u'_r r \left(C_{55}\omega'_x + C_{55}u'_r \right)$$

$$+ \delta u'_x r \left(C_{11}u'_x + C_{12}\frac{\omega}{r} + C_{13}\omega'_r \right) dxdr$$

$$= \int_\Omega \delta\omega r^2 \omega^2 \varrho dxdr \tag{5.17}$$

The radius-dependent stiffness matrix C and operator D are defined as:

$$D = \begin{bmatrix} 0 & 1 \\ \dfrac{\partial}{\partial x} & 0 \\ 0 & \dfrac{\partial}{\partial x} \\ \dfrac{\partial}{\partial r} & 0 \\ 0 & \dfrac{\partial}{\partial r} \end{bmatrix} \tag{5.18}$$

$$C(r) = \begin{bmatrix} \dfrac{1}{r}C_{22} & C_{12} & 0 & 0 & C_{23} \\ C_{12} & rC_{11} & 0 & 0 & rC_{13} \\ 0 & 0 & rC_{55} & rC_{55} & 0 \\ 0 & 0 & rC_{55} & rC_{55} & 0 \\ C_{23} & rC_{13} & 0 & 0 & rC_{33} \end{bmatrix} \tag{5.19}$$

The element equation can be defined as:

$$\int_\Omega (DN)^T C_{(r)} (DN) \, dx dr \, \tilde{u} = \int_\Omega N r^2 \omega^2 \varrho \, dx dr \tag{5.20}$$

In which the structure of the shape function matrix N is denoted by $\tilde{u}$. To obtain the optimal configuration of the flywheel with the most even stress distribution, the following function should be minimized.

$$O = \int_r (\sigma - \bar{\sigma})^2 \, dr = \min \tag{5.21}$$

In this equation, $\bar{\sigma}$ is the average stress. The square of stress difference should be minimized for constant mass and constant rotational energy. Each function is defined as:

$$2\pi\varrho \int_r \left(t_{(r)} - t_0 \right) dr = 0 \tag{5.22}$$

$$\omega^2 \pi\varrho \int_r r^3 \left(t_{(r)} - t_0 \right) dr = 0 \tag{5.23}$$

5.4 Future perspective

Development in the area of FES systems will continue via penetration of the technology into several more applications along with enhancements of FES technical features. The application of FES systems as energy storage systems is mature, but it does have drawbacks such as high capital cost and low energy density that still need to be addressed. Future investigations are expected to advance related technologies (such as material, electromagnetic, and mechanical engineering) such that the the angular velocity of FES systems could safely increase, while their costs decrease. The maximum reported velocity of the rotor is 100,000 rpm, and to increase this velocity much stronger materials are required. Thus, to give FES systems more penetration and future applications, investigations about new materials and composites should continue.

According to advancements in electrical machine technology, there has been increasing attention to industrial multiphase machines, which are complex, multi-variable electro-mechanical systems for variable velocity applications. This is due to these system's better fault tolerance and power-per-phase splitting characteristics compared to those of conventional three-phase machines. The future application of this type of machine will be an interesting alternative in FES systems. Consequently, in the future, research and applied works in this area will be guided towards these types of machines. In an FES system, the power electronic interface has a very important role. Using a conventional matrix converter as a power electronic interface results in decreased output gain of 86.6%. In future works, more efforts should be paid to introducing new topologies to increase this output gain. In addition, the cost of magnetic bearings is very high and they possess some nonlinear stabilities. Moreover, there is an interaction between the magnetic field of the magnetic bearing and that of the linear motor. This feature causes complexity and instability in the controlling system. It is expected that scientific efforts should be focused on the cost of magnetic bearings as well as improved and innovative methods to solve instability problems.

Overall, although there is still much to be done in the field of FES technology, this area is quite mature as it is now and these signs of progress will most likely lead to marginal improvements in the capacity and efficiency of these systems. Therefore, it is too optimistic to believe that FES technology can be a competitive alternative for large-scale MES systems such as pumped hydropower storage, compressed air energy storage system, and so on.

References

[1] A. Arabkoohsar, L. Machado, M. Farzaneh-Gord, R.N.N. Koury, Thermo-economic analysis and sizing of a PV plant equipped with a compressed air energy storage system. Renew. Energy 83 (2015), https://doi.org/10.1016/j.renene.2015.05.005.

[2] A. Arabkoohsar, Combination of air-based high-temperature heat and power storage system with an organic Rankine cycle for an improved electricity efficiency. Appl. Therm. Eng. 167 (2020) 114762, https://doi.org/10.1016/j.applthermaleng.2019.114762.

[3] S. Wicki, E.G. Hansen, Clean energy storage technology in the making: an innovation systems perspective on flywheel energy storage. J. Clean. Prod. 162 (2017) 1118–1134, https://doi.org/10.1016/j.jclepro.2017.05.132.

[4] R. Guerrero-Lemus, J.M. Martínez-Duart, Electricity storage BT—renewable energies and CO2. in: R. Guerrero-Lemus, J.M. Martínez-Duart (Eds.), Cost Analysis, Environmental Impacts and Technological Trends, 2012th ed., Springer, London, 2013, pp. 307–333, https://doi.org/10.1007/978-1-4471-4385-7_15.

[5] A. Arabkoohsar, Combined steam based high-temperature heat and power storage with an organic Rankine cycle, an efficient mechanical electricity storage technology. J. Clean. Prod. (2019) 119098, https://doi.org/10.1016/j.jclepro.2019.119098.

[6] P. Breeze, Chapter 6. Flywheels. in: P. Breeze (Ed.), Power System Energy Storage Technologies, Academic Press, 2018, pp. 53–59, https://doi.org/10.1016/B978-0-12-812902-9.00006-7.

[7] A. Soomro, M.E. Amiryar, K.R. Pullen, D. Nankoo, Comparison of performance and controlling schemes of synchronous and induction machines used in flywheel energy storage systems. Energy Procedia 151 (2018) 100–110, https://doi.org/10.1016/j.egypro.2018.09.034.

[8] P. Gambôa, J.F. Silva, S.F. Pinto, E. Margato, Input–output linearization and PI controllers for AC–AC matrix converter based dynamic voltage restorers with flywheel energy storage: a comparison. Electr. Power Syst. Res. 169 (2019) 214–228, https://doi.org/10.1016/j.epsr.2018.12.023.

[9] M. Rivera, P. Wheeler, A. Olloqui, D.A. Khaburi, A review of predictive control techniques for matrix converters—part I, in: 2016 7th Power Electronics, Drive Systems & Technologies Conference, IEEE, 2016, pp. 582–588.

[10] R. Östergård, Flywheel Energy Storage: A Conceptual Study, Uppsala University, 2011.

[11] H. Bernhoff, Magnetic bearings in kinetic energy storage systems for vehicular applications, J. Electr. Syst. 7 (2011) 225–236.

[12] A.S. Alsagri, A. Arabkoohsar, H.R. Rahbari, A.A. Alrobaian, Partial load operation analysis of trigeneration subcooled compressed air energy storage system. J. Clean. Prod. (2019) 117948, https://doi.org/10.1016/j.jclepro.2019.117948.

[13] H. Chen, T.N. Cong, W. Yang, C. Tan, Y. Li, Y. Ding, Progress in electrical energy storage system: a critical review, Progr. Nat. Sci. 19 (3) (2009) 291–312.

[14] R. Sebastián, R. Peña Alzola, Flywheel energy storage systems: review and simulation for an isolated wind power system. Renew. Sust. Energ. Rev. 16 (2012) 6803–6813, https://doi.org/10.1016/j.rser.2012.08.008.

[15] A. Emadi, A. Nasiri, S.B. Bekiarov, Uninterruptible Power Supplies and Active Filters, CRC Press, 2017.

[16] D.R. Brown, W.D. Chvala, Flywheel energy storage: an alternative to batteries for UPS systems, Energy Eng. 102 (2005) 7–26.

[17] A. Abazari, H. Monsef, B. Wu, Coordination strategies of distributed energy resources including FESS, DEG, FC and WTG in load frequency control (LFC) scheme of hybrid isolated micro-grid. Int. J. Electr. Power Energy Syst. 109 (2019) 535–547, https://doi.org/10.1016/j.ijepes.2019.02.029.

[18] L. Shen, Q. Cheng, Y. Cheng, L. Wei, Y. Wang, Hierarchical control of DC microgrid for photovoltaic EV charging station based on flywheel and battery energy storage

system. Electr. Power Syst. Res. 179 (2020) 106079, https://doi.org/10.1016/j.epsr.2019.106079.

[19] Hazle Township, Pennsylvania | Beacon Power.

[20] World's Largest Flywheel Energy Storage System, Beacon Power, LLC, https://beaconpower.com.

[21] A.R. Millner, A flywheel energy storage and conversion system for photo-voltaic applications. in: J. Silverman (Ed.), Energy Storage, Pergamon Press, 1980, pp. 494–508, https://doi.org/10.1016/B978-0-08-025471-5.50050-6. This work was sponsored by the United States Department of Energy.

[22] S. Ghosh, S. Kamalasadan, An integrated dynamic modeling and adaptive controller approach for flywheel augmented DFIG based wind system, IEEE Trans. Power Syst. 32 (2016) 2161–2171.

[23] M. Deepak, R.J. Abraham, F.M. Gonzalez-Longatt, D.M. Greenwood, H.-S. Rajamani, A novel approach to frequency support in a wind integrated power system. Renew. Energy 108 (2017) 194–206, https://doi.org/10.1016/j.renene.2017. 02.055.

[24] Technology–Flybrid Systems.

[25] GKN and the Go-Ahead Group using F1 technology to improve fuel efficiency of London buses | Automotive World.

[26] It's the New BROOKESbus! | Oxford Bus Company.

[27] A. Rupp, H. Baier, P. Mertiny, M. Secanell, Analysis of a flywheel energy storage system for light rail transit. Energy 107 (2016) 625–638, https://doi.org/10.1016/j.energy.2016.04.051.

[28] PUNCH Flybrid—PUNCH Flybrid Is the Leading Developer of Flywheel-Based Energy Storage Systems.

[29] E. Allison-Napolitano, Flywheel: Transformational Leadership Coaching for Sustainable Change, Corwin Press, 2013.

[30] P.R.S. Shelke, D.G. Dighole, A review paper on dual mass flywheel system, Int. J. Sci. Eng. Technol. Res. 5 (2016) 326–331.

[31] V. Babuska, S.M. Beatty, B.J. Deblonk, J.L. Fausz, A review of technology developments in flywheel attitude control and energy transmission systems, in: 2004 IEEE Aerospace Conference Proceedings (IEEE Cat. No.04TH8720), 2004, pp. 2784–2800.

[32] B. Bolund, H. Bernhoff, M. Leijon, Flywheel energy and power storage systems. Renew. Sust. Energ. Rev. 11 (2007) 235–258, https://doi.org/10.1016/j.rser.2005.01.004.

[33] D. Johnson, Design Considerations and Implementation of an Electromechanical Battery System, University of Cape Town, 2007.

[34] G. Genta, Spin tests on medium energy density flywheels, Composites 13 (1982) 38–46.

[35] Y. Bai, Q. Gao, H. Li, Y. Wu, M. Xuan, Design of composite flywheel rotor, Front. Mech. Eng. China 3 (2008) 288–292.

[36] Y. Bai, H. Li, Y. Wu, M. Xuan, Design of composite flywheel rotor, Opt. Precis. Eng. 15 (2007) 852–857.

[37] V. Kale, M. Secanell, A comparative study between optimal metal and composite rotors for flywheel energy storage systems. Energy Rep. 4 (2018) 576–585, https://doi.org/10.1016/j.egyr.2018.09.003.

[38] A.C. Arvin, C.E. Bakis, Optimal design of press-fitted filament wound composite flywheel rotors, Compos. Struct. 72 (2006) 47–57.

[39] E.L. Danfelt, S.A. Hewes, T.-W. Chou, Optimization of composite flywheel design, Int. J. Mech. Sci. 19 (1977) 69–78.

[40] C.H. Zweben, Comprehensive Composite Materials, Elsevier, 2000.

[41] G. Genta, The shape factor of composite material filament-wound flywheels, Composites 12 (1981) 129–134.

[42] S. Saidi, A. Djebli, Analysis of the effects of materials on the resistance of the flywheel. Procedia Manuf. 22 (2018) 675–682, https://doi.org/10.1016/j.promfg.2018.03.097.

[43] Y. Wang, X. Dai, K. Wei, X. Guo, Progressive failure behavior of composite flywheels stacked from annular plain profiling woven fabric for energy storage. Compos. Struct. 194 (2018) 377–387, https://doi.org/10.1016/j.compstruct.2018.04.036.

[44] Q. Chen, C. Li, Y. Tie, K. Liu, Progressive failure analysis of composite flywheel rotor based on progressive damage theory, J. Mech. Eng. 9 (2013).

[45] T. Changliang, Mechanical Analysis and Experimental Study of Commingled and Woven Composite Flywheel, Tsinghua University, 2013.

[46] G.R. Kress, Shape optimization of a flywheel, Struct. Multidiscip. Optim. 19 (2000) 74–81.

[47] J.M. Chern, W. Prager, Optimal design of rotating disk for given radial displacement of edge, J. Optim. Theory Appl. 6 (1970) 161–170.

[48] R. Okou, A.B. Sebitosi, P. Pillay, Flywheel rotor manufacture for rural energy storage in sub-Saharan Africa. Energy 36 (2011) 6138–6145, https://doi.org/10.1016/j.energy.2011.07.051.

[49] N.A. Dagnaes-Hansen, I.F. Santos, Magnetically suspended flywheel in gimbal mount—nonlinear modelling and simulation, J. Sound Vib. 432 (2018) 327–350.

[50] J.G. Bitterly, Flywheel technology: past, present, and 21st century projections, IEEE Aerosp. Electron. Syst. Mag. 13 (1998) 13–16.

[51] R. Pena-Alzola, D. Campos-Gaona, M. Ordonez, Control of flywheel energy storage systems as virtual synchronous machines for microgrids, in: 2015 IEEE 16th Workshop on Control and Modeling for Power Electronics, IEEE, 2015, pp. 1–7.

[52] J. Gonçalves de Oliveira, Power Control Systems in a Flywheel Based All-Electric Driveline, Uppsala University, 2011.

[53] X. Lyu, L. Di, S.Y. Yoon, Z. Lin, Y. Hu, A platform for analysis and control design: emulation of energy storage flywheels on a rotor-AMB test rig. Mechatronics 33 (2016) 146–160, https://doi.org/10.1016/j.mechatronics.2015.12.007.

[54] I. Hamzaoui, F. Bouchafaa, A. Talha, Advanced control for wind energy conversion systems with flywheel storage dedicated to improving the quality of energy. Int. J. Hydrog. Energy 41 (2016) 20832–20846, https://doi.org/10.1016/j.ijhydene.2016.06.249.

[55] N. Hiroshima, H. Hatta, M. Koyama, J. Yoshimura, Y. Nagura, K. Goto, Y. Kogo, Spin test of three-dimensional composite rotor for flywheel energy storage system. Compos. Struct. 136 (2016) 626–634, https://doi.org/10.1016/j.compstruct.2015.10.035.

[56] Y. Qiu, S. Jiang, Suppression of low-frequency vibration for rotor-bearing system of flywheel energy storage system. Mech. Syst. Signal Process. 121 (2019) 496–508, https://doi.org/10.1016/j.ymssp.2018.11.033.

[57] B. Xiang, W.O. Wong, Vibration characteristics analysis of magnetically suspended rotor in flywheel energy storage system. J. Sound Vib. 444 (2019) 235–247, https://doi.org/10.1016/j.jsv.2018.12.037.

[58] N.A. Dagnaes-Hansen, I.F. Santos, Magnetically suspended flywheel in gimbal mount—test bench design and experimental validation. J. Sound Vib. 448 (2019) 197–210, https://doi.org/10.1016/j.jsv.2019.01.023.

[59] R.K. Miyamoto, A. Goedtel, M.F. Castoldi, A proposal for the improvement of electrical energy quality by energy storage in flywheels applied to synchronized grid generator systems. Int. J. Electr. Power Energy Syst. 118 (2020) 105797, https://doi.org/10.1016/j.ijepes.2019.105797.

[60] Y. Miyazaki, K. Mizuno, T. Yamashita, M. Ogata, H. Hasegawa, K. Nagashima, S. Mukoyama, T. Matsuoka, K. Nakao, S. Horiuch, T. Maeda, H. Shimizu, Development of superconducting magnetic bearing for flywheel energy storage system.

Cryogenics (Guildf). 80 (2016) 234–237, https://doi.org/10.1016/j.cryogenics.2016.05.011.

[61] R. Hebner, J. Beno, A. Walls, Flywheel batteries come around again, IEEE Spectr. 39 (2002) 46–51.

[62] S. Sabihuddin, A.E. Kiprakis, M. Mueller, A numerical and graphical review of energy storage technologies, Energies 8 (2015) 172–216.

[63] M. Farhadi, O. Mohammed, Energy storage technologies for high-power applications, IEEE Trans. Ind. Appl. 52 (2015) 1953–1961.

[64] B.H. Kenny, P.E. Kascak, R. Jansen, T. Dever, W. Santiago, Control of a high-speed flywheel system for energy storage in space applications, IEEE Trans. Ind. Appl. 41 (2005) 1029–1038.

[65] R. Pena-Alzola, R. Sebastián, J. Quesada, A. Colmenar, Review of flywheel based energy storage systems, in: 2011 International Conference on Power Engineering, Energy and Electrical Drives, IEEE, 2011, pp. 1–6.

[66] Y. Yali, W. Yuanxi, S. Feng, The latest development of the motor/generator for the flywheel energy storage system, in: 2011 International Conference on Mechatronic Science, Electric Engineering and Computer, IEEE, 2011, pp. 1228–1232.

[67] Y. Yu, Y. Wang, G. Zhang, F. Sun, Analysis of the comprehensive physical field for a new flywheel energy storage motor/generator on ships, J. Mar. Sci. Appl. 11 (2012) 134–142.

[68] M.E. Amiryar, K.R. Pullen, D. Nankoo, Development of a high-fidelity model for an electrically driven energy storage flywheel suitable for small scale residential applications, Appl. Sci. 8 (2018) 453.

[69] S.J. Chapman, Electric Machinery Fundamentals, WCB/McGraw-Hill, 2004.

[70] M.A.M. Ramli, A. Hiendro, S. Twaha, Economic analysis of PV/diesel hybrid system with flywheel energy storage, Renew. Energy 78 (2015) 398–405, https://doi.org/10.1016/j.renene.2015.01.026.

[71] T.S. Davies, N. Larsen, A regenerative drive for incorporating flywheel energy storage into wind generation systems, in: Proceedings of the 24th Intersociety Energy Conversion Engineering Conference, IEEE, 1989, pp. 2065–2069.

[72] G.O. Suvire, P.E. Mercado, Active power control of a flywheel energy storage system for wind energy applications, IET Renew. Power Gener. 6 (2012) 9–16.

[73] A. Abdel-Khalik, A. Elserougi, A. Massoud, S. Ahmed, A power control strategy for flywheel doubly-fed induction machine storage system using artificial neural network, Electr. Power Syst. Res. 96 (2013) 267–276.

[74] R. Sebastián, R. Peña-Alzola, Control and simulation of a flywheel energy storage for a wind diesel power system, Int. J. Electr. Power Energy Syst. 64 (2015) 1049–1056.

[75] M.I. Daoud, A.S. Abdel-Khalik, A. Elserougi, A. Massoud, S. Ahmed, N.H. Abbasy, An artificial neural network based power control strategy of low-speed induction machine flywheel energy storage system, J. Adv. Inf. Technol. 4 (2013) 61–68.

[76] G. Genta, Kinetic Energy Storage: Theory and Practice of Advanced Flywheel Systems, Butterworth-Heinemann, 2014.

[77] S.R. Holm, Analytical modeling of a permanent-magnet synchronous machine in a flywheel, IEEE Trans. Magn. 43 (5) (2007) 1955–1967.

[78] A. Dhand, Design of Electric Vehicle Propulsion System Incorporating Flywheel Energy Storage, City University London, 2015.

CHAPTER SIX

Pumped heat storage system

Ahmad Arabkoohsar
Department of Energy Technology, Aalborg University, Esbjerg, Denmark

Abstract

A pumped thermal energy storage (PTES) system, also known as pump heat energy storage, is one of the most recent technologies introduced for electricity storage at large- and medium-scale applications. The PTES system, which is in the category of mechanical energy storage (MES) systems, is a promising technology that is likely to be broadly implemented worldwide in the near future. This system can be used not only for electricity storage/production but also for cogeneration of electricity and heat or even tri-generation of electricity, heat, and cold. This chapter presents a detailed description and fundamentals of the system and how it works. It also examines the state of the art of the technology, which is immature compared to other MES systems. Finally, it gives detailed mathematical modeling of the system and discusses the future perspectives of this technology.

6.1 Fundamentals

It is known for decades that a heat engine can be used for power generation utilizing the temperature difference between high-temperature and low-temperature heat sources [1]. As such, the same method might also be a smart option for electricity storage. However, the idea of storing electricity as high-temperature heat is quite nascent. This idea has been proposed because direct electricity storage on a large scale is quite difficult and technologies capable of doing this (e.g., compressed air) are either in the immature stages or suffer from technical, economic, or geographical restrictions [2]. This principle simply proposes the storage of surplus electricity in the form of high-temperature heat in a cheap yet efficient and stable storage medium so that it can be reclaimed to drive a conventional heat-driven power generation cycle whenever needed [3]. In these systems the heat storage medium is mainly gravel or rocks, and the power block is a Rankine cycle, Brayton cycle, or other. [4]. This system, which is called high-temperature heat and power storage, has been designed, analyzed, and demonstrated in Northern Europe in a variety of configurations. For example, the Siemens AG project and a Danish project led by SEAS-NVE are based on an integration of a large-scale packed bed of stones and a Rankine cycle [5]. This unit is built on a capacity of 35 MW in Hamburg with a fairly low roundtrip efficiency of 40% and is best suited for recycling the existing thermal power plants [6]. A combination of a large-scale packed bed of rocks and a multi-stage Bryton cycle was proposed and investigated by researchers in Denmark, resulting in a power-to-power efficiency of about 30% and a power-to-heat efficiency of about 50% [7]. Adding an organic Rankine cycle to these configurations for better electricity efficiencies (up to 45% and 36%, respectively) has also been investigated in [8, 9].

Parallel to the aforementioned works on high-temperature heat and power storage, the concept of the PTES system is also being developed. This system is not only based on the main idea of electricity storage in the form of heat storage in cheap and efficient materials (e.g., rock beds) but is also based on the same operation principle as heat engines working between a high-temperature and a low-temperature heat source [10]. Fig. 6.1 presents a schematic diagram of a PTES system. According to the figure, the system comprises one cold storage unit, a hot thermal storage unit, a compressor, a turbine, and a recuperator as the main elements. Both the heat and cold storage units are packed beds of stones in insulated tanks.

Like any other energy storage system, apart from the standby mode, there are two main operation stages for a PTES unit: charging and discharging. During the charging phase, a gas flow (in a closed loop) is compressed and heated through a compressor with a large enough pressure ratio. Here, the focus is more on heat generation at high temperatures (greater than 500°C) rather than high pressures, but the compressor unit should have the required pressure ratio (greater than 8 for the top pressure of up to

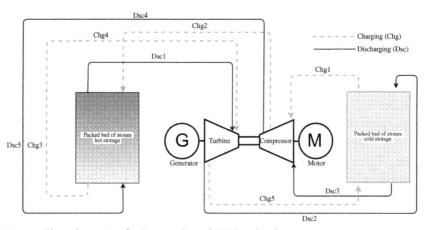

Fig. 6.1 The schematic of a Brayton-based PTES technology.

20 bar) to make the desired temperature level of the high-temperature source. The compressed gas flow then flows through the high-temperature packed stone bed carrying the heat to the packed bed, leaving the storage at a much lower temperature. In this way, the thermal energy of the gas flow is transferred to the heat storage unit and therefore the temperature of the packed bed increases. Here, it is very important that the heat storage unit be charged in a well-stratified manner so that the gas leaving from the hot storage unit is totally discharged in temperature. Therefore, the gas flow is so controlled that it facilitates the stratification of the tank. Next, the cold gas flow that is still at high pressure passes through a turbine to not only cover a portion of the required compressor work but also to produce gas at very low temperatures (way lower than $-100°C$) and low pressure. Finally, the subcooled and depressurized gas then flows into the cold storage unit. Here again, the storage unit should be well stratified; therefore, the gas flow is under rigorous control. The gas gives its cold energy to the storage tank and increases in temperature itself. After this stage, the gas at low pressure and low temperature (but not at the subcooled level) starts the loop again and enters the compressor to be compressed and heated up. Functionally, this is a heat pump cycle with a coefficient of performance of about 130%–150%, depending on the temperature ranges [11].

In contrast, during the discharging phase, the flow direction is reversed, resulting in a Brayton cycle similar to a gas turbine cycle in which the high-temperature storage acts as the combustion chamber for preheating the compressed gas flow before expansion. In this state, the cold gas flow from the cold storage unit flows through the compressor and is heated up to some extent. Then, it flows through the hot storage unit and reaches the maximum temperature level of the system. The hot yet high–pressure gas flow is expanded through the turbine, which is coupled to an electricity generator for power generation. The discharge efficiency is about 40%–45%, again depending on the temperature ranges. Considering the charging and discharging efficiencies of the system as well as the losses of the thermal storage units, the total thermodynamic roundtrip efficiency of the system is expected to be about 60%–65%. The suitable gases for a PTES system are argon, nitrogen, and others that can tolerate the severely high and low operating temperature ranges of the cycle. Air is also perfectly suitable for these pressures and temperature ranges. Another feature of the aforementioned monatomic gases (argon, nitrogen, etc.) distinguishing them from other stable gases like air is that they heat and cool much more strongly than air at the same pressure change levels and this can be extremely effective for reducing the costs of the system [11].

Understanding the operation principles of the system in the charging and discharging modes, the thermodynamic diagrams of the cycle can be plotted. For example, Fig. 6.2 shows the Pressure-Volume (P-V) of the working gas through the cycle of the PTES system during the charging and discharging processes, assuming isentropic compression and expansion. Similarly, the T-s diagrams and P-h diagrams can be plotted for a PTES cycle as well.

Naturally, for compressors and expanders, an isothermal process is optimal for achieving maximum efficiency. However, regardless of the fact that isothermal expansion and compression processes are not possible realistically, the PTES cycle operates based on the temperature difference of the cold and hot heat sources; thus, an isothermal process is not of interest here at all. As the compression and expansion processes take place so quickly, the processes approach being isentropic. However, for argon and hydrogen, when working over the design operating at high and low pressure ranges of the PTES system, an isentropic compression could result in a too-high temperature, which is not favorable for the system due to the restrictions in the materials' thermal strength. The same applies to the expansion process where too-low temperature is achieved over the isentropic expansion process. Therefore, the compression and expansion processes are tailored to happen via polytropic processes by modifying the fluid's properties to obtain the desired temperatures [12].

An important point here is that the above configuration shows distinguished turbine and compressor units (distinguished motor and generator, too) and two separate lines for the gas flow for the charging and discharging processes, while in many cases a PTES system is based on an inverse loop direction where the turbine of the charging process acts as the compressor

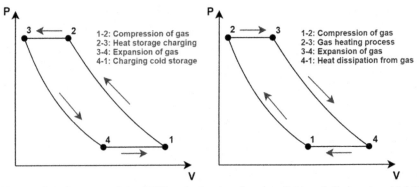

Fig. 6.2 P-V diagrams of the PTES cycle for the charging *(left)* and discharging *(right)* processes.

of the discharging process and vice versa. In this case, the motor should also act as a generator [12].

Another noteworthy matter is that a PTES system can be used for heat and cold supply. Thus, integration of the system to not only the electricity grid but also to the district heating and cooling systems would be possible by adding heat exchangers to the system. Fig. 6.3 presents a schematic of a PTES system that is reversible in the cycle direction and has interactions with both district heating and cooling systems.

In this configuration, a PTES system can be implemented at a fairly low cost and work at high efficiency, resulting in a lower unit cost of energy than in other conventional MES systems such as pumped hydropower energy storage. PTES is claimed to result in a decreased unit cost of energy less than $20 USD/MWh over a useful lifetime of 20 years [13].

PTES technology has several advantages, the most important being low cost and high energy storage density. In addition, a PTES system is not

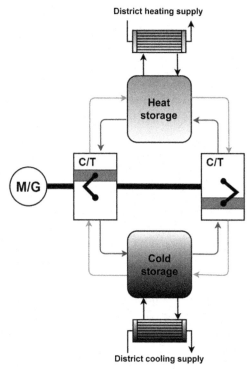

Fig. 6.3 The schematic of a PTES system in a reversible cycle format and interaction with district energy systems; *M/G*: motor-generator, *C/T*: compressor-turbine; *clockwise-arrows*: charging direction, *counterclockwise-arrows*: discharging directions.

restricted geographically, making it implementable everywhere. A PTES system can be constructed for a capacity range of 2–5 MW and large-scale storage capacities up to hundreds of megawatts. For gigawatt capacities, several units of PTES systems could be constructed at the same site. The agility of the system is a big plus for electricity storage as it needs to go to standby, charging, or discharging modes frequently, especially when operating as an ancillary service for the electricity market. The potential for the trigeneration of power, heat, and cold makes this system a great asset for future energy systems where there is a strong need for integration of energy sectors [14].

However, the electricity-to-electricity storage of the system (which is about 60%) is not that impressive, which is an important drawback. The expected efficiency of the PTES system is not only way lower than that of batteries (with efficiencies greater than 90% [15]) but also lower in comparison with other available MES systems. For example, the roundtrip efficiency of a pumped hydropower energy storage unit (for which the output is only power) is greater than 80% [16]. For compressed air energy storage systems, especially advanced models, the power generation efficiency can be greater than 70% and even up to near 80% [17, 18]. Greater electricity storage efficiency, rather than a high overall energy efficiency when considering cold and heat productions, is important as it has been shown that the electricity sector will be the main energy sector among all in the future, even playing a key role in supporting heating and cooling grids [19]. Additionally, PTES systems are a young technology and there is still much to be learned. Therefore, the not well-developed literature and poor state of the art of this technology are the major drawbacks. Naturally, the technology will not get the chance for real-life implementation until it is sufficiently developed for successful pilot-scale demonstration.

In addition to the previously discussed design, which is by far the most important and promising design of PTES systems, there are also other energy storage systems that work based on the same principles. Two of these are "transcritical Rankine PTES" and "compressed heat energy storage." The former system is an organic Rankine cycle with CO2 as its working fluid coupled with cold and hot storage units. Fig. 6.4 shows the schematic of a transcritical Rankine PTES unit for the two operation modes of charging and discharging [20].

According to the figure, the system in the charging mode acts as a heat pump in which CO_2 is the main working fluid, and in the discharging mode operates as an organic Rankine cycle. This system is claimed to reach

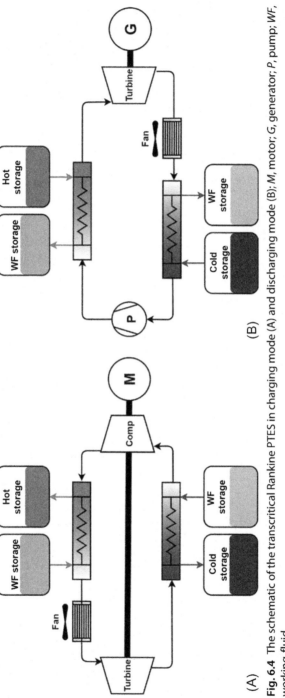

Fig. 6.4 The schematic of the transcritical Rankine PTES in charging mode (A) and discharging mode (B); *M*, motor; *G*, generator; *P*, pump; *WF*, working fluid.

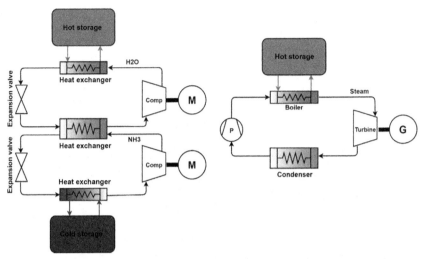

Fig. 6.5 The schematic of the compressed heat energy storage system in charging mode *(left)* and discharging mode *(right)*.

roundtrip efficiency of 60% provided that all the processes and components are well optimized [21].

Fig. 6.5 illustrates the schematic of a compressed heat energy storage system. In the charging mode, the system is a combination of two heat pump cycles in a series arrangement, one with ammonia as its working fluid at lower temperature and the other with water as the working fluid at higher temperature. In the discharging mode, the system will is a conventional Rankine cycle with water/steam as the working fluid. A detailed explanation of how this system operates and its advantages and disadvantages is given in Ref. [22].

6.2 State-of-the-art

Compared to other well-known energy storage technologies such as batteries, pumped hydropower storage, compressed air energy storage, flywheel storage, and so on, the PTES system is quite new. Therefore, its literature is still immature not only in practice but also in terms of research and development activities.

The basic design of this technology, based on the scheme discussed in this chapter (i.e., the Brayton cycle concept), was first proposed by Desrues et al. [23] in 2010, without calling it a PTES (or pumped heat energy storage) system. They explained how this system takes advantage of the artificially created large temperature difference between the two thermal storage units in the charging mode for power generation in the discharging mode. They also presented the basic thermodynamic model of the system and developed a numerical model simulating the performance of the system when operating. This primary work on this technology also considered argon as the main working fluid of the cycle where the maximum temperature and pressure are assumed to reach 1000°C and 4.6 bar, while the minimum temperature is −70°C. Based on these temperature levels and considered characteristics, the efficiency of the system was found to be about 67%. Naturally, this high-temperature level for storage is not practically feasible due to the material restrictions, but it was a great start to develop this concept. Then, in 2012, the first article experimentally investigating this technology and specifically mentioning the title of "pumped heat electricity storage" was published. In this work, Howes [24] experimentally tested a PTES system (in different prototypes) with a reciprocating turbine and compressor, a packed bed of gravel, and argon as the working fluid. This system was designed for the working fluid with a flow rate of 12.56 kg/s, maximum temperature and pressure of 500°C and 12.13 bar (after the compression stage), and minimum pressure and temperature of −166°C and 1 bar (after the expansion process).

Following these preliminary works, the number of research studies on PTES technology increased. White et al. [25] presented a thermodynamic (energy and exergy) analysis of the PTES system, focusing on the energy density, power density, and roundtrip efficiency of the system and investigating the parameters affecting these features between thermal storage units with 500°C and −150°C temperatures. The main finding of this work was the great dependence of the system efficiency on the exergy losses and irreversibilities in the turbomachinery of the cycle. Thess [26] developed the basic thermodynamic model of PTES and created a simple equation that could estimate the efficiency of the system as a function of the thermal storage unit temperature. The main finding of this study was that a PTES system with a thermal storage unit temperature of greater than 400°C could compete with conventional compressed air energy storage systems in terms of roundtrip efficiency. Further, by increasing the temperature of the storage, the efficiency of the system became competitive with with that of advanced adiabatic compressed air energy storage systems. In another work, White et al. [27] specifically investigated (both analytically and numerically) the thermal wave propagation in the packed bed of stones as the main mechanism of cold and heat storage in PTES systems. They concluded that nonlinear thermal waves take place in cold fronts, which could lead to shock-like feature formation, and that exergy losses associated with heat transfer processes within the pack beds are quite high and result in 15% of the thermal energy. McTigue et al. [28] did a parametric study of PTES systems and tried to optimize the system performance with a certain focus on the thermal energy storage units of the system. In this work, the sensitivity of parametric study results to a number of effective parameters in the cycle was assessed and the optimization results presented as trade-off surfaces for cycle energy efficiency as well as power and energy densities. It was shown that doing the optimizations, the effect of pressure drop through the storage units and the irreversibilities associated with them on the overall energy efficiency of the system will be quite low, and a roundtrip efficiency of 70% is not impossible for the system. Ni and Caram [29] used the exponential matrix approach to analyze the relation of the roundtrip efficiency of the PTES system to other parameters such as the characteristics of turbomachinery, heat transfer resistance, and others. It was shown that the roundtrip efficiency, the stored energy per unit heat capacity, and the storage bed utilization factor are the main parameters that can best describe the performance of the system.

Guo et al. [30] did a performance analysis and optimization of PTES technology and compared it with a pumped hydropower energy storage system. They identified the main irreversibility sources in the PTES system, determined the logical ranges of technical parameters in the cycle, and proposed the main criteria for optimum design of the PTES (and other energy storage systems). In another work, Guo et al. [31] analyzed PTES system performance and tried to develop a function relating to the roundtrip efficiency of the system to its rate of power production. They also investigated the impacts of the main control parameters of the PTES system on its technical performance and discussed the optimal range of these parameters. Smallbone et al. [32] compared the levelized cost of storage achievable for the PTES system with those reported for other grid-scale electricity storage technologies and concluded that PTES can surely compete with pumped hydropower energy storage technology economically with the further advantage of not being geographically restricted. For the PTES system, the levelized cost of storage was found to be in the range of 8.9–11.4 cents Euro/kWh. Roskosch and Atakan [33] analyzed thermodynamically a scheme of PTES in which there was a combination with a Rankine cycle. The results of this study indicated that the PTES roundtrip efficiency will increase (up to 70%) as the storage temperature increases. Regarding the organic Rankine cycle, it was found that the efficiency of the cycle strongly depends on the turbine inlet condition at which, for higher efficiency, the superheating should be restricted as much as possible. Laughlin [34] analyzed a PTES system in which there were four storage tanks (two hot and two cold units), the thermal energy storage material was molten salt at the high-pressure side and hydrocarbon liquist at the low-pressure side. They analyzed the system thermodynamically, discussed the limitations of the materials (including thermal storages) for the PTES systems, calculated the cost per stored energy and cost per engine capacity for two different working fluids of nitrogen and argon, and concluded that the roundtrip efficiency of the system can be as good as that of a pumped hydropower system.

Benato [35] presented a performance and economic investigation of a novel design of a PTES system in which an electrical heater is used before the hot storage to increase the temperature of the compressed air so that the maximum temperature of the system (and thereby, the performance of the system) is not affected by the compressor pressure ratio changes (as a result of operating load changes) as it was in the other configurations of the PTES. In addition, a heat exchanger was used to maintain the turbine inlet temperature at the design level, which led to an easily regulatable maximum cycle temperature just in accordance with the energy amount to be stored. This work generated the mathematical model of the system and assessed the system performance and cost for five different thermal storage materials of two different shapes. In another work, Benato and Stoppato [36] continued their research on their proposed scheme of PTES investigating the appropriate heat transfer fluid and thermal storage materials (nine storage materials and two working fluids) and testing different control strategies and reservoirs' discretization. The conclusion was that the roundtrip efficiency of this PTES scheme is not as high as others previously proposed and investigated; however, considering real-life turbomachinery already available in the market, the energy density range of $70–430\,kWh/m^3$ of thermal storage and the specific cost of $50–180$ Euro/kWh in different designs could be achieved for this system.

In 2018, Benato and Stoppato [36], after giving an overview of electricity storage technologies, presented a detailed literature survey of PTES technology in different designs and configurations. Georgiou et al. [37] presented a techno-economic analysis of PTES systems and compared that with the liquid-air electricity storage system. The conclusion was that the PTES has, of course, a lower technology readiness level as it is a newer concept, but presents a better roundtrip efficiency making it more competitive at greater electricity spot prices. The levelized cost of storage for the liquid-air energy system was found to be less than that of the PTES system in this study. Chen et al. [38] presented a thermodynamic analysis on a modified version of PTES in which an electrical heater was used to increase the temperature of the gas and thereby increase the storage capacity and efficiency of the cycle. Calling this system a high-temperature PTES system, a combination of it with an organic Rankine cycle (in five different configurations) for further performance improvement was also investigated. Two different gases (air and argon) were assessed as the working fluids of the PTES system. Air was selected as the main working fluid due to economic and thermal performance considerations. It is found out that combining PTES and an organic Rankine cycle for waste heat recovery significantly improves the roundtrip efficiency of the system. Among all the possibilities, the most promising solution is when the high-temperature PTES comes in a parallel arrangement with the organic Rankine cycle unit, offering an energy storage density as high as $219 \, MJ/m^3$.

Wang et al. [39] studied the effects of an unbalanced mass flow rate of the working fluid through the thermal storage units of the PTES system on the performance of the system. It was shown that the unbalanced flow proportions for the hot and cold storages are 0.62% and 0.26%, respectively, while for the closed PTES cycle, the unbalanced mass flow rate is 0.36%. In addition, a sensitivity analysis on the effects of porosity, pressure ratio, and the thermal capacity of the storage material on the unbalanced mass flow rate and roundtrip efficiency of the system was carried out. In another work, Wang et al. [40] proposed the transient analysis approach for analyzing the performance of a PTES system with 10 MW and 4-h charging capacity. They discussed the roundtrip efficiency and power output stability of the system under transient operation and the effects of dimensions of the storage as well as the size of the stone particles, polytropic efficiency, and pressure ratio of turbomachinery on these parameters. In this work, it was, surprisingly, concluded that helium is a better working fluid than argon for PTES systems. Steinmann et al. [41] investigated the possibility of a Rankine-based PTES system (i.e., compressed heat energy storage concept) for sector-integration (between electricity and heat sectors) via low-temperature heat integration (e.g., a low-temperature geothermal heat source, a seasonal solar-powered energy storage system, etc.). In this system, the maximum temperature of the cycle is lower than in conventional compressed heat energy storage design, resulting in smoothening the material requirements for the charging cycle of the system. The system would be able to cogenerate low-temperature heat (for district heating purposes) and electricity.

Roskosch et al. [42] analyzed the general thermodynamic potential and limits of the PTES system in compressed heat storage design, investigating the relation of the power output and the roundtrip efficiency of the system as well as the impacts of the turbomachinery isentropic efficiencies on system efficiency. Optimal but hypothetical working fluids were considered for the system. It was concluded that the roundtrip efficiency of the system decreases as the maximum storage temperature increases, and the effect of expander isentropic efficiency in the organic Rankine cycle is quite significant on the roundtrip efficiency of the whole system. Frate et al. [43] presented a multi-criteria analysis of an improved design of a Rankine-based PTES system in which waste heat is recovered. In this work, the energy and exergy efficiencies of the cycle, energy density, and energy ratio were quantified and a trade-off between these parameters was investigated. It was found out the PTES system with the heat recovery unit exploiting heat at a temperature lower than 80°C is better than the system without any regenerator loop and can result in a roundtrip energy efficiency of 55% and energy storage density of $15\,kWh/m^3$. It was also found that greater energy efficiency will happen at a lower exergy efficiency so that increasing the exergy efficiency of the system reduces the roundtrip efficiency to as low as 40%. Georgiou et al. [44] made a detailed comparison of the PTES system with liquid-air energy storage technology in low-carbon electricity systems. It was found that location and capacity are the important factors to achieve acceptable system values and costs for both of these technologies. The PTES system will have a higher system value than the other system anyway. However, the PTES system will have greater power and capital costs compared to the liquid-air energy storage, making it less attractive (in terms of economic aspects) at smaller plant capacities. It is finally concluded that although the complicated perspective of energy systems and the decarbonization challenges make it difficult to confidently say which of these two systems will be more broadly implemented in the future, the clear fact is that both PTES and liquid-air energy storage technologies are promising and can play significant roles in the future of renewable-based energy systems.

Apart from the previously discussed research, a group of researchers in England, at the Sir Joseph Swan Centre for Energy Research at Newcastle University, in cooperation with Isentropic Limited, took the first steps towards pushing the state of practice of this technology. They started to build the first pilot-scale PTES unit with a capacity of 150 kW input power and a storage capacity of 600 kWh [13]. The aim is to develop and demonstrate PTES system units with 2–5 MW capacities. This project includes the development of efficient and practical facilities for the key components of the system including novel engine pistons, valves, storage units, and so on [13].

6.3 Mathematical model

In this section, we present a step-by-step formulation of the PTES technology (in the Brayton-based design) by which a detailed energy analysis of the system will be possible. The mathematical model of the system comes in a component-by-component format and then the performance evaluation criteria of the integrated system are presented.

The first component to be modeled is the compressor. For a compressor, based on an isentropic process, the temperature output of the compressor, which is a determining parameter on the work consumption of the compressor, is calculated as [45]:

$$T_2 = T_1 \left(\frac{P_2}{P_1}\right)^{\frac{k-1}{k}} \tag{6.1}$$

In which T is the temperature of the fluid through the compressor in K, P is the pressure of the gas in kPa, and the indices 2 and 1 represent the outlet and inlet conditions of the compressor, respectively. Also, k is the thermal capacity ratio of the given gas, which is used when the process is isentropic. It was, however, mentioned that an isentropic process would lead to a too-high output temperature when argon, for example, is used in the system (considering the higher and lower pressure ranges required in the system). Therefore, the compression is managed to take place in a polytropic process. In this case, the preceding formula is rewritten as:

$$T_2 = T_1 \left(\frac{P_2}{P_1}\right)^{\frac{n-1}{n}} \tag{6.2}$$

Where n is the polytropic factor of the process, which should be lower than the value of k for the gas.

Having the temperature output of the compressor, one could calculate the enthalpy of the gas leaving the compressor and thereby the rate of work of the compressor through the compression process as:

$$\dot{W}_{comp} = \dot{m}(h_2 - h_1) = \frac{\dot{m}nRT_1}{n-1}\left[\left(\frac{P_2}{P_1}\right)^{\frac{n-1}{n}} - 1\right] \tag{6.3}$$

Where $\dot{m}$ is the flow rate of the gas through the compressor and h represents specific enthalpy of the gas.

After leaving the compressor, the high-temperature pressurized gas passes through the high-temperature thermal storage unit. The heat storage unit is to be a packed bed of gravels. Fig. 6.6 shows the schematic of a packed bed of gravel in a cylindrical shape as a heat storage unit with the gas flow entering and exiting into/from that.

In this case, a variety of models could be used, but one of the simplest yet most accurate methods is the one-dimensional two-phase Schumann model [46]. In this method, it is assumed that the temperature gradient of the packed bed of gravel is only in the height direction (the direction of the compressed gas flow through the storage and temperature stratification) and there is a uniform temperature distribution in other directions. Thus,

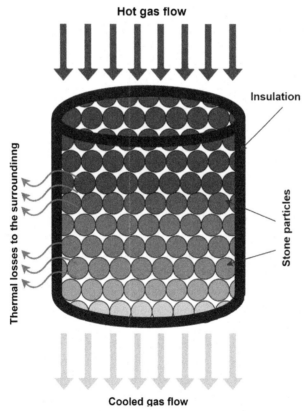

Fig. 6.6 The schematic diagram of a cylindrical shape packed bed of gravels with the up-down gas flow through it.

the heat transfer model will be one-dimensional. The model is a two-phase model as it generates individual temperature distribution profiles for the gas flow and the stone particles. Based on this mode, the temperature distribution equations of the stone particles and the gas flow will be as [46]:

$$\rho_s c_s (1 - \epsilon) \frac{dT_s}{dt} = h_v (T_a - T_s) + \overline{k} \frac{d^2 T_s}{dx^2} \tag{6.4}$$

$$\rho_g c_{p,g} \epsilon \frac{dT_a}{dt} = h_v (T_s - T_g) - \frac{\dot{m}_g c_{p,g}}{A} \frac{dT_g}{dx} - \frac{D\pi U}{A} (T_g - T_{sur}) \tag{6.5}$$

In which ρ and c_p represent the density and heat capacity, $\dot{m}$ and ϵ are the mass flow rate of the gas flow and the porosity ratio of the storage, and A and D are the cross-sectional area and the diameter of the storage (considering the storage as a vertical cylinder). The subscripts g, s, and sur represent the gas flow, the stone particles, and the surrounding of the storage, respectively. Finally, h_v, $\overline{k}$, and U are the volumetric heat transfer coefficient, the effective heat conduction coefficient, and the overall heat loss coefficient, respectively, given by [46]:

$$h_v = 700 \left[\frac{\dot{m}_g}{Ad} \right]^{0.76} \tag{6.6}$$

$$\overline{k} = k_s (1 - \epsilon) + k_g \epsilon \tag{6.7}$$

$$U = \left(\frac{1}{\overline{h}_{in}} + \frac{R}{k_{ins}} \ln \left(\frac{R + \delta_{ins}}{R} \right) + \frac{R}{k_s} \ln \left(\frac{R + \delta_{ins} + \overline{R}}{R + \delta_{ins}} \right) \right)^{-1} \tag{6.8}$$

In the last three equations, d and k are the equivalent diameter of the stone particles and the thermal conduction coefficient, and $\overline{h}$ is the heat convection coefficient. Also, δ, R, and $\overline{R}$ are the thickness of the insulation, the radius of the storage, and the storage's thermal influential distance.

By the above set of formulas, the packed bed of stones could be modeled and its temperature distribution along the height calculated. In addition, the temperature of the gas coming out of the storage for entering the turbine could also calculated.

For the turbine, as it decreases the temperature of the gas, there should not be any problem with an isentropic process (i.e., lower temperature will not be a serious issue for the system materials, unlike the higher temperature for which there are technical restrictions). However, the general method for making a polytropic process in the compressor is adding some liquid particles in the gas flow. This will, therefore, affect the expansion process as well, resulting again in a polytropic expansion process. Having the same equation as Eq. (6.2), one could calculate the temperature outlet of the turbine. Thus, for calculating the power production rate of the turbine, which will be used for covering a portion of the compressor work, one has [47]:

$$\dot{W}_{turb} = \dot{m}(h_2 - h_1) = \frac{\dot{m}nRT_1}{n-1}\left(1 - \frac{P_2}{P_1}^{\frac{n-1}{n}}\right) \tag{6.9}$$

In which, again, 1 and 2 refer to the inlet and outlet conditions of the turbine, respectively.

The flow coming out of the turbine is at very low-temperature levels (e.g., $-160°C$ for the specific PTES case designed by [12]). Then, this sub-cooled flow goes through the second packed bed of stones, which is a cold storage. The same formulation as those presented for the hot storage may apply here. The only difference is that here the direction of the gas flow within the storage is from bottom to top.

Finally, having the rates of heat and cold energy stored in the thermal storage units, the rate of work consumption of the compressor, and the production of the turbine, one could make the following energy balance for the charging mode of the system:

$$\dot{E}_{surp} = \frac{\left(|\dot{W}_{comp}| - \dot{W}_{turb}\right)}{\eta_{mec}\eta_m} \tag{6.10}$$

Where $\dot{E}_{surp}$ is the rate of surplus power to be stored and η_{mec} is the mechanical efficiency of the couplings, and η_m is the electrical efficiency of the motor driving the compressor.

In the discharging mode, as the main components of the system are the same as in the charging mode operation, the same sort of formulas should be used for modeling the system. Here, naturally, there is no surplus power to charge the compressor. Rather, the turbine is connected to the electricity generator to supply power. The rate of power generation in the system could be calculated as:

$$\dot{E}_{supl} = \eta_{mec}\eta_g\left(\dot{W}_{turb} - |\dot{W}_{comp}|\right) \tag{6.11}$$

In which $\dot{E}_{supl}$ is the rate of power generation of the system and η_g is the efficiency of the generator.

In case of supplying energy for heating and cooling applications (e.g., to the district heating and cooling networks), there should be two heat exchangers in the system as well. For the heat exchangers in the cold and hot sides, assuming district heating and cooling supply, the inlet/outlet temperatures of the secondary fluid, which is pressurized water, will be 40°C/80°C and 12–15°C/6–8°C, respectively [48, 49]. For each of these heat exchangers, the following energy balance equation applies:

$$\dot{Q}_{pbs} = \dot{Q}_{des} = \dot{m}_{des}c_p\left(T_{sup} - T_{ret}\right) \tag{6.12}$$

Where $\dot{Q}_{pbs}$ is the rate of heat/cold energy withdrawn from the packed bed of stones, $\dot{Q}_{des}$ is the rate of energy supplied to the district energy systems, and T_{sup} and T_{ret} are the supply and return temperatures of the secondary fluid to the district energy system.

Having the inlet and outlet of the secondary side of the heat exchanger fixed, depending on the amount of energy to be supplied for the district energy systems, one should regulate the mass flow rate of working fluids in the primary and secondary sides. For this, the effectiveness factor of the heat exchangers should be known. The effectiveness factor of a counter-flow heat exchanger is defined as [50]:

$$\varepsilon = \frac{1 - e^{-NTU(1-C_r)}}{1 - C_r e^{-NTU(1-C_r)}} \tag{6.13}$$

Where NTU is the number of transfer units and C_r is the ratio of the thermal capacity of the primary and secondary fluids. These two parameters are given by:

$$NTU = {}^{UA}\!\big/\!{}_{c_{min}} \tag{6.14}$$

$$C_r = {}^{C_{min}}\!\big/\!{}_{C_{max}} \tag{6.15}$$

Where UA refers to the overall heat transfer coefficient of the heat exchangers. C_{min} and C_{max} are, respectively, the lower and the higher values of $\dot{m} \times c_p$ (mass flow rate multiplied by the specific thermal capacity at constant pressure) of the primary and secondary fluids.

Then, the amount of heat being transferred through heat exchangers can be reformulated in the following manner to calculate the temperatures and mass flow rate of the primary side of the heat exchanger (i.e., the thermal storage units' side):

$$\dot{Q}_{pbs} = \dot{m}_{pbs}c_{p,pbs}\left(T_{in,pbs} - T_{out,pbs}\right) = \varepsilon C_{min}\left(T_{in,pbs} - T_{in,des}\right) \tag{6.16}$$

Where $\dot{m}_{pbs}$ is the flow rate of the primary heat transfer fluid (the one transferring thermal energy from the thermal storage unit to the heat exchangers),

$c_{p, pbs}$ refers to its specific thermal capacity, and the subscripts $T_{in, pbs}$ and $T_{out, pbs}$ represent the inlet and outlet temperatures of the primary fluid into/from the heat exchangers.

Finally, having all the components and processes of the PTES system modeled, one could define the following overall energy performance indices:

$$\eta_{el} = \frac{\sum_{t=1}^{dsct} \dot{E}_{supl}}{\sum_{t=1}^{cgt} \dot{E}_{surp}} \qquad (6.17)$$

$$\eta_{egy} = \frac{\sum_{t=1}^{dsct} \left(\dot{E}_{supl} + \dot{Q}_{cold} + \dot{Q}_{heat} \right)}{\sum_{t=1}^{cgt} \dot{E}_{surp}} \qquad (6.18)$$

The former equation is the roundtrip efficiency of the PTES system when it is only appointed for electricity production and the latter considers all the heat, cold, and electricity as the output of the system. In these equations, *dsct* is the number of discharging time-steps of the system and *cgt* refers to the number of charging time-steps.

6.4 Future perspective

PTES is a cost-effective electricity storage technology with several important advantages that give it great potential among all MES systems for large-scale and broad use in future renewable-based energy systems globally. There is still, however, a lot to be resolved for reaching that point as the technology suffers from some disadvantages, too. The most important drawback of this system is its immature state of the art. For this, the literature needs more works in this context to determine the optimal design and specifications of the system. Dynamic modeling of the technology under fluctuating operating conditions is an important task for determining the performance of the system under intensely varying conditions, off-design operation, startup and load shifting, and so on. Then, laboratory work is are needed to support the results of dynamic simulation works in this context. Techno-economic analyses would be helpful for doing minor and major improvements in the system performance and achieving better cost rates and efficiencies. Finding the best operation strategies for the PTES system as operates in different configurations (i.e., power-only production, cogeneration, or trigeneration) in various types of power plants and different energy markets could be highly beneficial for pushing the state of the art of the system. Although all of this has been addressed to some extent, more extensive work needs to be done before pilot-scale demonstration. A pilot demonstration is, in fact, the first practical phase of the commercialization of the system. If the performance of the system in real-life operation and on the pilot scale is as expected through the simulations, labratory experiments, and dynamic modelings, the system will have the chance to be broadly implemented.

In addition to its immatureness, another big weakness of PTES technology is the existing uncertainty associated with the components of the system to do their tasks as smoothly and efficiently as expected for the appropriate working fluids (argon, nitrogen, etc.). For this technology, the compressor and turbine components (impellers and housings) are available as off-the-shelf components and are used in compressor-turbine combinations in large turbochargers and gas turbines, but not specifically for the monoatomic gases mentioned. In addition, the maximum allowable operating temperature and the efficiencies of the compressor and expander are of key importance to the overall roundtrip efficiency of the system. Elevating the temperature from 500°C (a typical maximum for industrial equipment) to 600°C (possible with advanced materials) will improve the roundtrip efficiency from about 60% to 70%. Improving the compressor efficiency from 85% (a typical maximum for open impellers) to 90% (a likely maximum for a custom-designed, shrouded impeller) will further improve the roundtrip efficiency from 70% to 75%. Also, packed stone beds have been studied extensively, but their use in PTES has novel elements that provide optimization potentials subject to further research. Packed beds are typically implemented as open–circuit, atmospheric-pressure units, with connection to turbomachinery through heat exchangers. Due to unavoidable thermal gradients, such heat exchangers invariably lead to irreversible loss of thermodynamic efficiency. The PTES system does not have heat exchangers in the primary circuit, but instead operates with a closed circuit and pressurized storage tanks. To facilitate the use of normal steel types in the tanks, they should be insulated internally, ensuring that tanks are maintained at roughly ambient temperature.

Having said this, one should note that for future works on PTES systems, there is a need to create new yet solid knowledge in the fields of:

1. thermal and mechanical design of appropriate yet optimized compressors (and expanders) compatible with the most efficient working fluids for the system,
2. optimization of PTES for the selection of the system parameters to achieve the optimal combination of efficiency and cost,
3. thermal design of an optimized packed stone bed or any alternative solutions for heat and cold storage,
4. dynamic modeling of the system and its performance under realistic off-design and partial-load operation conditions, and
5. testing the system on a small scale and validating simulation results.

In this way, the path will be paved for pilot-scale demonstration of PTES systems and thus open new windows for commercialization and broad implementation of this technology.

References

[1] J.P. Stark, Fundamentals of classical thermodynamics (Van Wylen, Gordon J.; Sonntag, Richard E.). J. Chem. Educ. 43 (1966) A472, https://doi.org/10.1021/ed043pA472.1.
[2] A. Arabkoohsar, L. Machado, M. Farzaneh-Gord, R.N.N. Koury, Thermo-economic analysis and sizing of a PV plant equipped with a compressed air energy storage system. Renew. Energy 83 (2015) https://doi.org/10.1016/j.renene.2015.05.005.
[3] A. Arabkoohsar, G.B.B. Andresen, Design and analysis of the novel concept of high temperature heat and power storage. Energy 126 (2017) 21–33, https://doi.org/10.1016/j.energy.2017.03.001.
[4] A. Arabkoohsar, G.B. Andresen, Thermodynamics and economic performance comparison of three high-temperature hot rock cavern based energy storage concepts. Energy 132 (2017) https://doi.org/10.1016/j.energy.2017.05.071.
[5] Siemens, Siemens High Temeprature Heat and Power Storage Project, https://www.siemens.com/press/en/pressrelease/?press=/en/pressrelease/2016/windpower-renewables/pr2016090419wpen.htm&content=WP, 2016.
[6] Greentech Media, Siemens Gamesa Starts Building Hot Rock Plant for Long-Duration Grid Storage, https://www.greentechmedia.com/articles/read/siemens-gamesa-starts-on-giant-thermal-storage-plant, 2017. (Accessed 5 December 2019).
[7] A. Arabkoohsar, G.B. Andresen, Dynamic energy, exergy and market modeling of a high temperature heat and power storage system. Energy 126 (2017) https://doi.org/10.1016/j.energy.2017.03.065.
[8] A. Arabkoohsar, Combined steam based high-temperature heat and power storage with an organic rankine cycle, an efficient mechanical electricity storage technology. J. Clean. Prod. 119098 (2019) https://doi.org/10.1016/j.jclepro.2019.119098.
[9] A. Arabkoohsar, Combination of air-based high-temperature heat and power storage system with an organic Rankine cycle for an improved electricity efficiency. Appl. Therm. Eng. 167 (2020) 114762, https://doi.org/10.1016/j.applthermaleng.2019.114762.
[10] Sinovoltaics–Zero Risk Solar™, Pumped Heat Electrical Storage, https://sinovoltaics.com/learning-center/storage/pumped-heat-electrical-storage/, 2019. (Accessed 5 April 2020).

[11] Energy Storage Association, Pumped Heat Electrical Storage (PHES), https://energystorage.org/why-energy-storage/technologies/pumped-heat-electrical-storage-phes/, 2020. (Accessed 5 April 2020).

[12] D. McKicley, Pumped Heat Energy Storage, The University of Edinburgh, 2016.

[13] Isentropic-Press Office-Newcastle University, https://www.ncl.ac.uk/press/articles/archive/2017/11/isentropic/, 2017. (Accessed 5 April 2020).

[14] A. Arabkoohsar, G.B. Andresen, Design and optimization of a novel system for trigeneration. Energy 168 (2019) 247–260, https://doi.org/10.1016/j.energy.2018.11.086.

[15] A. Poullikkas, A comparative overview of large-scale battery systems for electricity storage. Renew. Sust. Energ. Rev. 27 (2013) 778–788, https://doi.org/10.1016/j.rser.2013.07.017.

[16] J.I. Pérez-Díaz, M. Chazarra, J. García-González, G. Cavazzini, A. Stoppato, Trends and challenges in the operation of pumped-storage hydropower plants. Renew. Sust. Energ. Rev. 44 (2015) 767–784, https://doi.org/10.1016/j.rser.2015.01.029.

[17] A. Arabkoohsar, L. Machado, M. Farzaneh-Gord, R.N.N. Koury, The first and second law analysis of a grid connected photovoltaic plant equipped with a compressed air energy storage unit. Energy 87 (2015) 520–539, https://doi.org/10.1016/j.energy.2015.05.008.

[18] A. Arabkoohsar, L. Machado, R.N.N. Koury, Operation analysis of a photovoltaic plant integrated with a compressed air energy storage system and a city gate station. Energy 98 (2016) 78–91, https://doi.org/10.1016/j.energy.2016.01.023.

[19] H. Nami, A. Arabkoohsar, Improving the power share of waste-driven CHP plants via parallelization with a small-scale Rankine cycle, a thermodynamic analysis. Energy (2019) 27–36, https://doi.org/10.1016/j.energy.2018.12.168 In Press.

[20] M. Morandin, F. Maréchal, M. Mercangöz, F. Buchter, Conceptual design of a thermo-electrical energy storage system based on heat integration of thermodynamic cycles—part A: methodology and base case. Energy 45 (2012) 375–385, https://doi.org/10.1016/j.energy.2012.03.031.

[21] M. Morandin, M. Mercangöz, J. Hemrle, F. Maréchal, D. Favrat, Thermoeconomic design optimization of a thermo-electric energy storage system based on transcritical CO2 cycles. Energy 58 (2013) 571–587, https://doi.org/10.1016/j.energy.2013.05.038.

[22] W.D. Steinmann, The CHEST (compressed heat energy storage) concept for facility scale thermo mechanical energy storage. Energy 69 (2014) 543–552, https://doi.org/10.1016/j.energy.2014.03.049.

[23] T. Desrues, J. Ruer, P. Marty, J.F. Fourmigué, A thermal energy storage process for large scale electric applications. Appl. Therm. Eng. 30 (2010) 425–432, https://doi.org/10.1016/j.applthermaleng.2009.10.002.

[24] J. Howes, Concept and development of a pumped heat electricity storage device. Proc. IEEE 100 (2012) 493–503, https://doi.org/10.1109/JPROC.2011.2174529.

[25] A. White, G. Parks, C.N. Markides, Thermodynamic analysis of pumped thermal electricity storage. Appl. Therm. Eng. 53 (2013) 291–298, https://doi.org/10.1016/j.applthermaleng.2012.03.030.

[26] A. Thess, Thermodynamic efficiency of pumped heat electricity storage. Phys. Rev. Lett. 111 (2013) 110602, https://doi.org/10.1103/PhysRevLett.111.110602.

[27] A. White, J. McTigue, C. Markides, Wave propagation and thermodynamic losses in packed-bed thermal reservoirs for energy storage. Appl. Energy 130 (2014) 648–657, https://doi.org/10.1016/j.apenergy.2014.02.071.

[28] J.D. McTigue, A.J. White, C.N. Markides, Parametric studies and optimisation of pumped thermal electricity storage. Appl. Energy 137 (2015) 800–811, https://doi.org/10.1016/j.apenergy.2014.08.039.

[29] F. Ni, H.S. Caram, Analysis of pumped heat electricity storage process using exponential matrix solutions. Appl. Therm. Eng. 84 (2015) 34–44, https://doi.org/10.1016/j.applthermaleng.2015.02.046.

[30] J. Guo, L. Cai, J. Chen, Y. Zhou, Performance optimization and comparison of pumped thermal and pumped cryogenic electricity storage systems. Energy 106 (2016) 260–269, https://doi.org/10.1016/j.energy.2016.03.053.

[31] J. Guo, L. Cai, J. Chen, Y. Zhou, Performance evaluation and parametric choice criteria of a Brayton pumped thermal electricity storage system. Energy 113 (2016) 693–701, https://doi.org/10.1016/j.energy.2016.07.080.

[32] A. Smallbone, V. Jülch, R. Wardle, A.P. Roskilly, Levelised cost of storage for pumped heat energy storage in comparison with other energy storage technologies. Energy Convers. Manag. 152 (2017) 221–228, https://doi.org/10.1016/j.enconman.2017.09.047.

[33] D. Roskosch, B. Atakan, Pumped heat electricity storage: potential analysis and orc requirements. Energy Procedia 129 (2017) 1026–1033, https://doi.org/10.1016/j.egypro.2017.09.235.

[34] R.B. Laughlin, Pumped thermal grid storage with heat exchange. J. Renew. Sustain. Energy 9 (2017) 044103. https://doi.org/10.1063/1.4994054.

[35] A. Benato, Performance and cost evaluation of an innovative pumped thermal electricity storage power system. Energy 138 (2017) 419–436, https://doi.org/10.1016/j.energy.2017.07.066.

[36] A. Benato, A. Stoppato, Pumped thermal electricity storage: a technology overview. Therm. Sci. Eng. Prog. 6 (2018) 301–315, https://doi.org/10.1016/j.tsep.2018.01.017.

[37] S. Georgiou, N. Shah, C.N. Markides, A thermo-economic analysis and comparison of pumped-thermal and liquid-air electricity storage systems. Appl. Energy 226 (2018) 1119–1133, https://doi.org/10.1016/j.apenergy.2018.04.128.

[38] L.X. Chen, P. Hu, P.P. Zhao, M.N. Xie, F.X. Wang, Thermodynamic analysis of a high temperature pumped thermal electricity storage (HT-PTES) integrated with a parallel organic Rankine cycle (ORC). Energy Convers. Manag. 177 (2018) 150–160, https://doi.org/10.1016/j.enconman.2018.09.049.

[39] L. Wang, X. Lin, L. Chai, L. Peng, D. Yu, J. Liu, et al., Unbalanced mass flow rate of packed bed thermal energy storage and its influence on the joule-Brayton based pumped thermal electricity storage. Energy Convers. Manag. 185 (2019) 593–602, https://doi.org/10.1016/j.enconman.2019.02.022.

[40] L. Wang, X. Lin, L. Chai, L. Peng, D. Yu, H. Chen, Cyclic transient behavior of the Joule–Brayton based pumped heat electricity storage: modeling and analysis. Renew. Sust. Energ. Rev. 111 (2019) 523–534, https://doi.org/10.1016/j.rser.2019.03.056.

[41] W.D. Steinmann, D. Bauer, H. Jockenhöfer, M. Johnson, Pumped thermal energy storage (PTES) as smart sector-coupling technology for heat and electricity. Energy 183 (2019) 185–190, https://doi.org/10.1016/j.energy.2019.06.058.

[42] D. Roskosch, V. Venzik, B. Atakan, Potential analysis of pumped heat electricity storages regarding thermodynamic efficiency. Renew. Energy 147 (2020) 2865–2873, https://doi.org/10.1016/j.renene.2018.09.023.

[43] G.F. Frate, L. Ferrari, U. Desideri, Multi-criteria investigation of a pumped thermal electricity storage (PTES) system with thermal integration and sensible heat storage. Energy Convers. Manag. 208 (2020) 112530, https://doi.org/10.1016/j.enconman.2020.112530.

[44] S. Georgiou, M. Aunedi, G. Strbac, C.N. Markides, On the value of liquid-air and pumped-thermal electricity storage systems in low-carbon electricity systems. Energy 193 (2020) 116680, https://doi.org/10.1016/j.energy.2019.116680.

[45] A. Bejan, Advanced Engineering Thermodynamics, John Wiley & Sons, 2016.

[46] K. Attonaty, P. Stouffs, J. Pouvreau, J. Oriol, A. Deydier, Thermodynamic analysis of a 200 MWh electricity storage system based on high temperature thermal energy storage. Energy 172 (2019) 1132–1143, https://doi.org/10.1016/j.energy.2019.01.153.

[47] M.J. Moran, H.N. Shapiro, D.D. Boettner, M.B. Bailey, Principles of Engineering Thermodynamics, eighth ed., John Wiley & Sons, 2015.

[48] A. Arabkoohsar, M. Khosravi, A.S. Alsagri, CFD analysis of triple-pipes for a district heating system with two simultaneous supply temperatures. Int. J. Heat Mass Transf. 141 (2019) 432–443, https://doi.org/10.1016/j.ijheatmasstransfer.2019.06.101.

[49] A. Arabkoohsar, M. Sadi, A solar PTC powered absorption chiller design for co-supply of district heating and cooling systems in Denmark. Energy 193 (2020) https://doi.org/10.1016/j.energy.2019.116789.

[50] F.P. Incropera, T.L. Bergman, A.S. Lavine, D.P. DeWitt, Fundamentals of Heat and Mass Transfer. (2011) https://doi.org/10.1073/pnas.0703993104.

CHAPTER SEVEN

New emerging energy storage systems

Ahmad Arabkoohsar
Department of Energy Technology, Aalborg University, Esbjerg, Denmark

Abstract

After extensive discussion of various already known mechanical energy storage (MES) technologies in the previous chapters, this chapter gives an introduction to a number of new emerging electricity storage systems in this category. Although there are several newly introduced technologies and the number is increasing, this chapter is centered around the three most promising systems: subcooled compressed air energy storage (SCAES), which is also known as trigeneration compressed air energy storage (TCAES); high-temperature heat and power storage (HTHPS), which comes in the two forms of air-based and steam-based; and gravity energy storage (GES), which comes in a variety of designs. After introducing the technologies and reviewing their state of the art, we present the mathematical models required for analyzing these systems and discuss future perspectives.

7.1 Introduction of the technologies

7.1.1 Subcooled compressed air energy storage (SCAES) technology

In Chapter 3, we introduced compressed air energy storage (CAES) technology and its different configurations. As explained there, CAES, in all configurations, is one of the most promising mechanical energy storage (MES) systems due its acceptable roundtrip efficiency and fairly low cost as compared other existing MES technologies [1]. A CAES unit is likely to be seen in different configurations, which were also discussed in Chapter 3. The most important and well-known configurations of CAES include diabatic CAES [2], adiabatic CAES [3], isothermal CAES [4], and low-temperature CAES [5, 6].

The best roundtrip electrical efficiency expected from a CAES design is around 80% for an advanced isothermal CAES system, whereas in the simpler designs, such as diabatic CAES, the expected efficiency is around 30% [7]. Advanced isothermal CAES results in the greatest efficiency among all the possible designs. It uses multi-stage compressors and air turbines as well as thermal energy storage (TES) units to collect and store the heat produced in the compression process and use it to preheat the compressed air stream before each of the expansion stages. Thus, it reduces the required excess heat of the system and increases the efficiency [8]. Diabatic CAES, on the other hand, as the oldest version of CAES configurations, may use single or multi-stage compressors and air turbines, but does not take advantage of the heat flow produced in the compression stage and simply wastes it. Therefore, a diabatic CAES is less efficient than an isothermal one [9]. Adiabatic CAES is an improvement of diabatic CAES and uses thermal storage systems to utilize the heat generation potential of the system [10]. Finally, a low-temperature CAES system is practically an isothermal CAES that does not employ auxiliary heaters for reaching high temperatures before the expansion process. It only utilizes the heat potential of the CAES system itself (and probably a medium-temperature renewable heating system, e.g., solar thermal unit) to reach a temperature in the range of 250–300°C [11].

Apart from these CAES designs, the new concept of subcooled compressed air energy storage (SCAES) was recently introduced to the literature [12]. Based on the design strategy of this concept, an SCAES system not only removes the need for any auxiliary heaters for the expansion process but also does not offer any preheating process for the compressed air before being expanded through the air turbines. In this case, the collected heat during the compression stage (the charging process) can be used for other heating applications. The most straightforward method of using the generated heat flow is district heating in which the required temperature is about 80°C, while the return temperature is about 40°C [13].

On the other hand, not heating the airflow before the expansion results in too-low temperatures of the turbine outlet airflows, which means the generation of a massive amount of cold requiring special air expander technologies [14]. The outlet airflow temperature of the turbines depends on the pressure ratio of turbine stages and the inlet airflow temperature; however, considering ambient temperature as the inlet temperature and an expansion ratio in the range of three to five for each of the expansion stages, the outlet temperature will be in the range of −50 to −100°C. This is why this design is referred to as subcooled CAES [15]. Fig. 7.1 presents the schematic design of a double-stage compression/expansion SCAES system.

According to the figure, and compatible with the explanation of this technology, the SCAES system looks much like a multi-stage isothermal CAES system with the only difference being that there is no heating process before any of the expansion stages. The heat exchangers after any of the

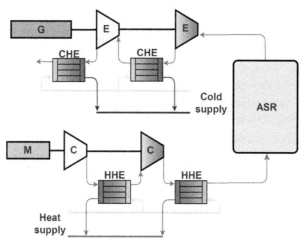

Fig. 7.1 A double-stage SCAES system configuration; *M*, motor; *C*, compressor; *HHE*, hot heat exchangers; *ASR*, air storage reservoir; *E*, expander; *CHE*, cold heat exchanger; *G*, generator.

expansion stages have been allocated to gather the generated cold potential through the turbines. Therefore, the system, besides its power production, will generate cold as it discharges. The cold production of the system can be used for any cold supply application, but similar to the heat production of the charging phase, the most straightforward method of using this cold flow is supplying it to the local district cooling grid, if any. For a typical district cooling supply system, the supply and return temperatures are 8°C and 15°C, respectively [16].

Having all these individual efficiencies, a SCAES unit offers a roundtrip heat production efficiency of about 80%–90% and roundtrip electricity and cold efficiencies of around 20%–35% [17]. This results in an overall energy efficiency (or coefficient of performance) of between 120% and 160%. Although the low power efficiency of the system compared to other MES systems is a drawback, its heat, cold, and power production show an overall energy efficiency (or the coefficient of performance) of about 1.2–1.6. This high overall efficiency and the feature of multi-generation it offers makes this system one of the most appropriate energy storage technologies for locations with high penetration of renewable energy as well as deployed district cooling, heating, and electricity networks such as Denmark, Norway, Sweden, and others [18]. In addition, according to Lund et al. [19], being a multi-generation system is highly beneficial as the system integrate different energy sectors, which is vital for future smart energy systems.

CAES technology was first introduced and modeled for the local electricity market of Denmark, in which there is a high penetration of wind power, by Arabkoohsar et al. [12] in 2019. An extensive thermodynamic analysis of an SCAES unit combined with a large-scale absorption chiller was presented in [15, 17]. Later, Arabkoohsar et al. [20] presented an analysis of the partial load operation of this technology, finding the system very adequate in terms of general energy production efficiency even in very low operating loads, though the power efficiency could collapse in such conditions. The combination of an SCAES unit with an organic Rankine cycle for the sake of improving the power efficiency of the system was proposed and investigated by Alsagri et al. [21], improving the system electrical efficiency by 20%.

The range of efficiencies expected from the system is a function of the number of compressor and expander stages as well as its design pressure ratios. However, these factors also figure into the cost of the system. Therefore, there is a need for a comprehensive techno–economic analysis of the costs and technical characteristics of the system's components.

7.1.2 High-temperature heat and power storage technology

High-temperature heat and power storage (HTHPS) is another MES system that was introduced in recent years [22]. The main idea of this technology is storing electricity in the form of heat in a TES unit and then using the stored heat for co-generation of heat and electricity in a conventional manner like a combined heat and power (CHP) system [23]. Choice of TES unit depends on the required temperature of the system and the availability of materials. However, studies of this system so far have only focused on packed beds of rocks as a cheap and reliable method of heat storage at high temperatures [24].

Based on the designs of this system so far, in the charging mode, the electrical energy is converted to high-temperature heat flow at almost 100% efficiency via electrical coils. The generated heat is transferred to the packed bed of rocks by circulating airflow through the electrical coils and then top to bottom through the packed bed of stones via fans. Fig. 7.2 presents an illustration of the packed bed of stones in an HTHPS system. As shown, an electrical coil is used to transfer the electricity flow to the TES unit by employing a number of fans circulating air from the bottom to the electrical coils at the top of the storage and from there into the packed bed of stones.

The methods proposed for reclaiming the stored heat for the co-generation of heat and electricity (i.e., the power block), however,

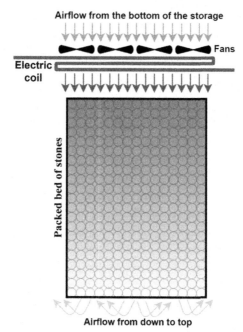

Fig. 7.2 Illustration of a packed bed of rocks and how it is used to store electricity at high-temperature heat form in an HTHPS system [25].

classify the HTHPS technologies into two main categories: air-based and steam-based [26]. We discuss these two designs in the sections that follow.

7.1.2.1 Air-based HTHPS

Fig. 7.3 presents the configuration of an air-based HTHPS system. According to the figure, the power block of the system is a double-stage

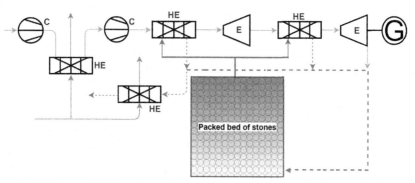

Fig. 7.3 Air-based HTHPS system with air and water flows through the heat exchangers and turbomachinery.

compression/expansion Bryton cycle with cooling and heating heat exchangers to increase the efficiency of the cycle. Similar to the SCAES system, a larger number of stages increases efficiency, but also increases the cost of the system. Thus, a detailed and rigorous techno–economic trade-off is needed to determine the most appropriate and cost-effective configuration of the power production unit [17]. Arabkoohsar et al. [27] proposed a system with triple compression/expansion stages and a pressure ratio of 3 as an optimal solution.

For the above system, the charging mode operation has already been discussed (the packed bed of stones is in operation and the power block is in standby mode). In the discharging phase, both the power block and the packed bed of stones will be working. At this stage, the compressors generate heat intake and compress the ambient air. There are heat exchangers after each stage of the compressor to cool down the airflow before entering the next stage compressor and thereby enhance the efficiency of the cycle. Here, of course, as the system has only two compressor stages, there is only one intercooler in the system. This heat exchanger will collect the generated heat in the airflow and use it for a heat supply application. Most of the studies conducted on HTHPS have proposed this heat be supplied to district heating systems with pressurized water as the main heat carrier (which will be the secondary working fluid of the intercooler, too) [28].

Then, after the second compressor, the compressed air is heated up to the desired temperature before the expansion process. This heating process is carried out via another heat exchanger, which is supplied by the high-temperature heat stored in the packed bed of stones. As there are two expansion stages, two preheating heat exchangers would also be needed here. The heated airflows coming out of these heat exchangers are expanded through the expanders for producing work. The discharged airflows from these two heat exchangers and that of the last turbine stage (as it has still a high temperature) are taken back to the packed bed of stones to slow down the rate of discharge. Since the pressure of the packed bed should not change over time, the surplus airflow taken into the system is used for further supply of the district heating system by the use of another heat exchanger. Taking this all into account, one can see that the described system generates work by two expanders (and then this work is converted to electricity by a generator) and supplies heat by two heat exchangers (one is the intercooler of the compressor stages and the other is the heat exchanger taking advantage of the surplus air taken into the system while operating). This configuration, depending on the number of compression and expansion stages and the pressure ratio of turbomachinery, may offer electrical, heat, and overall efficiencies up to 30%, 60%, and 90%, respectively [22].

7.1.2.2 Steam-based HTHPS

Fig. 7.4 illustrates the schematic of a steam–based HTHPS system with a Rankine cycle in a conventional design with a triple-stage steam turbine and two regenerators. In this configuration, unlike the air-based system in which the hot air coming from the packed bed of rocks is used for preheating the air stream before the expanders, the hot air supplied by the packed bed of stones is used to vaporize the water stream pressurized by the pump, generating high-pressure, high-temperature superheat steam. This steam is expanded through a triple-stage turbine between which there are two regeneration loops to increase the efficiency of the cycle. The condensed steam after the last stage of the low-pressure turbine goes to the condenser for recovery. Here, as the objective is to co-generate heat and electricity, the temperature of the last stage of the turbine is kept high enough to be able to supply the rejected heat of the steam flow through the condenser to the local district heating system. This means less electricity production of the cycle and not taking perfect advantage of the steam flow through the turbines [29]. The supply temperature of the existing district heating systems is 80°C, while this is to change to lower temperature levels of 50–55°C for low-temperature district heating systems [30] or 35–40°C in ultralow-temperature district heating systems [30]. Regardless, the condensed steam is then pumped into the boiler for repeating the cycle and co-generating heat and electricity. The electricity efficiency of this cycle is about 35%, while its

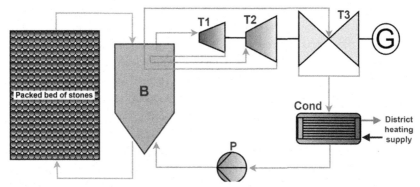

Fig. 7.4 Schematic diagram of a steam-based HTHPS system; *B*, boiler; *T1*, high-pressure turbine; *T2*, medium-pressure turbine; *T3*, low-pressure turbine; *Cond*, condenser; *P*, pump.

heat efficiency can be as high as 60%, resulting in overall energy efficiency of about 95% in a roundtrip charging-discharging cycle. This scheme of the HTHPS system (i.e., the Rankine-based design) has recently been demonstrated at the pilot scale in Germany [31].

Comparing the two HTHPS designs, one can conclude that an air-based design is more cost-effective (owing to its simple design) and offers the key advantage of agility (meaning fast response to sudden changes in the demand and availability of power). This agility is one of the most important features of an electricity storage system. On the other hand, the steam-based design presents a greater overall efficiency as well as better heat production and electricity efficiency, but it is not agile due to the long start-up time required for the steam to be prepared for work generation. The cost of construction of a steam-based HTHPS system is much greater than that of an air-based system [26].

7.1.3 Gravity energy storage technology

A gravity energy storage (GES) system is a unit that uses the force of gravity as the medium for storing electricity. In other words, a GES system stores electricity in the form of a heavy weight taken to higher elevations. When discharging, the weight is released to move down, actuating an electricity generator for producing power. Unlike the previous two MES technologies introduced, for which the scientific literature presents a number of articles covering various aspects of their performance and characteristics, GES technology does not yet have a rich literature. The main reason for this is that storing energy using gravity force is a fairly novel idea, proposed in the last couple of years, and therefore not many researchers and companies with related activities know about it or believe in it yet. The few articles available in the literature and other online sources, including the websites of the companies behind the technology and some online videos, introduce a few different designs of GES technology and provide information about the fundamentals of operation of these systems. According to these sources, GES technology might come in different configurations, either on the ground or under seawater. The latter class of GES technologies are called ocean-GES systems. Due to the personal opinion of the author about the promising future of this technology, these systems are introduced and discussed in the sections that follow.

Fig. 7.5 presents an illustration of a ground-GES system. According to the figure, a super large and heavy weight (a rock piston) is located

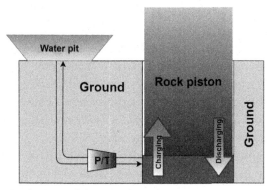

Fig. 7.5 Schematic of a ground-GES system.

vertically in a deep hole in the ground. In addition, there is a large enough source of water beside this vertical hole (e.g., a natural lake or an artificially constructed water pit). For storing electricity in the charging phase of the system, electricity is used to drive some pumps to produce a high-pressure water flow that is pumped below the rock piston. In this way, as the charging process continuous, the weight comes higher and higher. A higher position of the piston means a larger amount of electricity stored. In contrast with this process, during the discharging phase (i.e., when the GES unit needs to supply electricity to the grid), the piston is released to push the pressurized water below it through hydraulic turbines (mainly the pumps that are now reversed in operation) to generate rotational work. This rotational work is utilized with an electricity generator to produce power. The expanded water flow then goes to its primary origin or source. Such a system is, indeed, another version of pumped hydropower energy storage (PHES), which does not have the limitation of geographical needs for building up huge water dams. The efficiency of this system is expected to go beyond pumped hydropower, as the source of the pressure of the system is mainly the super-heavy rock piston that keeps the pressure behind the hydraulic turbines almost constant over the entire course, from the top position to the bottom of the hole. In reality, the pressure, of course, changes a little as the salted seawater below the piston and behind the turbine creates 1 bar pressure for every approximately 10 m of its height. Based on a rule of thumb, one could say that the potential energy released from a piston's up-down travel (which is to be converted to electricity via the turbo-generator package) would be about 10 kJ for any meter of motion if the piston has a weight of 1000 kg. Therefore, for having the storage capacity in large scales (e.g., the order of a few hundred MWs and a storage duration of a few days or so), the piston needs to be very large. Here, of course, the construction of the hole and the piston is not too costly as the materials of the system are natural and freely available. Having an available water source so that there is no need for an artificially created dam considerably lowers the cost of the system. The efficiency of this system is expected to be in the same order as (or even slightly greater than) a PHES system, which is in the range of 80%–90%.

The other possibility of a ground–GES system is not using hydraulic power to lift the piston but rather using a motor. Then, for the discharging mode, the weight is released, dropping it down, and actuates an electricity generator (which is the motor in reverse operation mode). Fig. 7.6 illustrates this system and how it works as an energy storage system [32].

Also, Fig. 7.7 shows the schematic of a GES system under seawater or ocean–GES. As its name suggest, an ocean–GES is implemented in the sea/ocean bed. In this system, there are a number of hanging heavy weights at almost the sea surface level. These weights are also connected to long cables so that if they fall down towards the sea bottom, their motion can rotate a shaft around which the cable is twisted. The shaft is also connected to an electricity generator to convert to rotational work to electricity. Knowing the structure of the system, one could explain the two charging and discharging mode of operation of this system as follows:

- In the charging mode, all (or some) of the weights that are down at the sea bottom are connected to the cables by the use of a robot. The cable moves upward by a motor (the generator works as a motor in a reverse

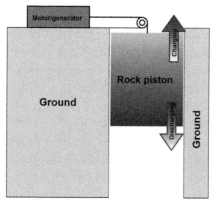

Fig. 7.6 Another configuration of ground-source GES.

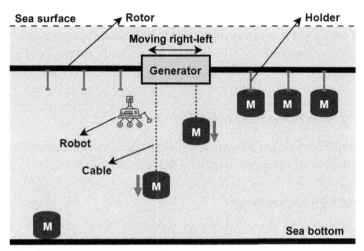

Fig. 7.7 Configuration of an ocean-GES system.

operation direction). The energy received by the motor is, indeed, the electricity that is to be stored. In fact, in this way, electricity is stored in the form of potential energy under seawater. As the surplus electricity to be stored continues, more and more weights could be taken up, meaning more energy is being stored.

• In the discharging mode, as there is a need for electricity to be stored by the energy storage system, the robot connects the cables to the weights and releases them from the holders. Thus, the weights move downward and actuate the electricity generator to produce power.

Given this information, one could say that ocean-GES technology is a reverse case of PHES in which instead of concrete dams retaining water, the system uses the limitless source of ocean water to retain some heavy weights (most likely concrete weights as the best choice) with floats. There, however, in case of implementing the system in the deep ocean, the system can take advantage of extremely higher elevation differences (e.g., about 3000–4000 m).

This system is applicable where the seawater depth is quite big because there is a direct and strong relationship between the amount of producible/storable power and the distance that the weights travel downwards/upwards in a discharging/charging mode. Clearly, similar to a ground-GES system, here also there is a strong relation between the capacity of the system and the mass of the weights so that the heavier the weights are, the higher the storage capacity of the system. The feasibility of this system is, of course, a matter of how long the cables need to be to transmit the produced power to the grid, as deep ocean water is not readily accessible in many locations.

Other configurations of GES technology including ocean- and ground-GES designs are found in Ref. [33].

7.2 Mathematical model

In this section, we present the mathematical model required to do an energy analysis of the aforementioned MES technologies.

7.2.1 SCAES technology

An SCAES system, like other energy storage technologies, works in two operating phases of charging and discharging. Therefore, the energy analysis models for the two operation phases are presented separately.

In the charging phase, when excess electricity is to be stored, the model governing the system operation is similar to that of the advanced adiabatic CAES system. Therefore, the electricity to be stored is employed to run a multi-stage compressor. For such a compressor, having the subscripts c and a referring to the compressor (with n stages) and the airflow, one can write [34]:

$$\dot{W}_c = \sum_{j=1}^{n} (\dot{m}_a w_c)_j \tag{7.1}$$

Where $\dot{W}_c$ is the total work to be used by the compressor in all stages, which is to be equal to the electricity supplied for the storage (neglecting the small mechanical losses of the compressor). Also, the parameter $\dot{m}_a$ refers to the mass flow rate of the compressed air that each of the compressor stages could generate, while w_c represents compressor stages' specific work consumption. The latter might be simply calculated by a basic thermodynamics formula related to adiabatic compressor work.

With r_c as the pressure ratio of compressor stages, k as the specific heat ratio of air, and $\eta_{is,\,c}$ as the compressor stages' isentropic efficiency, the following equation calculates at what temperature the compressed air stream will exit each of the compressor stages:

$$T_e = T_i \left(1 + \frac{r_c^{\left(\frac{k-1}{k}\right)} - 1}{\eta_{is,c}} \right) \tag{7.2}$$

According to Fig. 7.1, there will be a heat exchanger after each compressor stage (i.e., the intercooler) to collect the heat generated through the compressor stage. For a better heat collection performance, counter-flow plate heat exchangers are suggested to be employed here. For such a heat exchanger, with an effectiveness factor of ε, the rate of heat transfer between the fluids can be calculated by:

$$\dot{Q}_h = \varepsilon C_{min}\left(T_{a,i} - T_{f,i}\right) = \dot{m}_a\left(c_{p,i}T_{a,i} - c_{p,e}T_{a,e}\right) \tag{7.3}$$

Where C_{min} and C_{max} are the extensive heat capacities of the air stream and the secondary working fluid through the heat exchanger, respectively. Extensive heat capacity is defined as the mass flow rate of the fluid multiplied by its heat capacity. T is temperature and the indices i and e are related to the inlet and exit conditions of the fluids into/out of the heat exchanger. The inlet temperatures of the airflow and the working fluid stream (indicated by f) into the heat exchangers are both known and the exit temperatures are calculated by the following two correlations, respectively:

$$T_{a,e} = \frac{c_{p,i}T_{a,i} - \dot{Q}_{hx}\big/_{\dot{m}_a}}{c_{p,e}} \tag{7.4}$$

$$T_{f,e} = \frac{\dot{Q}_{hx}\big/_{\dot{m}_f} + c_{f,i}T_{f,i}}{c_{f,e}} \tag{7.5}$$

The effectiveness factor of the intercooler, since it is going to be a gas-fluid heat exchanger, will be quite high. According to Ref. [35], a gas-fluid heat exchanger can present an effectiveness factor of up to 0.95. The effectiveness factor is calculated as:

$$\varepsilon = \frac{1 - \exp\left(-NTU\left(1 - \frac{C_{min}}{C_{max}}\right)\right)}{1 - C_r\exp\left(-NTU\left(1 - \frac{C_{min}}{C_{max}}\right)\right)}; \text{ where}: NTU = \frac{UA}{C_{min}} \tag{7.6}$$

Where U and A are the overall heat transfer coefficient of the heat exchanger and the heat transfer area.

Calculating the mass flow rate and temperature of the air being compressed through the compressor stages, the pressure of the air storage chamber (represented by the subscript asr) at any moment over the charging process of the SCAES system can be calculated by:

$$P^{\lambda}_{asr} = \left(\frac{m_{asr}RT_{asr}}{V_{asr}}\right)^{\lambda} \text{ where } :m^{\lambda}_{asr} = m^{\lambda-1}_{asr} + \dot{m}_a \text{ and } T_{asr} = T_s \qquad (7.7)$$

Here, the parameters m_{asr} and V_{asr} are the mass of the air within the storage vessel and its volume, R is the gas constant of air, and T_s is the temperature of the storage walls, which is assumed to be equal to its surroundings. Also, the superscript λ is the time-step counter.

The preceding formulation gives the energy balance of the SCAES system in the charging phase. For the discharging process, the main parameter is the amount of power that should be produced based on the demand of the network or the power plant connected to the energy storage unit ($\dot{E}_d$). The required work to be converted to electricity should be provided by the air expanders, which is given by:

$$\dot{W}_t = \frac{\dot{E}_d}{\eta_g} = \sum_{j=1}^{n} (\dot{m}_a w_t)_j \qquad (7.8)$$

In which η_g is the energy conversion efficiency of the electricity generator (which may be in the range of 0.95–0.98). The rate of work production of each turbine stage is indicated by w_t in the above correlation, which can be calculated by:

$$w_t = RT_i \frac{\mu}{\mu-1} \left(\frac{1 - r_t^{\left(\frac{k-1}{k}\right)}}{\eta_{is,t}}\right) \qquad (7.9)$$

Here, r_t refers to the pressure ratio of the air expander and $\eta_{is,\ t}$ represents its isentropic efficiency. As mentioned, for an SCAES system, regular air expanders cannot be used. Rather, air expanders suitable for too-low temperatures must be employed. According to Ref. [36], such an expander (e.g., a single screw expander) can offer an isentropic efficiency of up to 0.65.

Here again, there will be a heat exchanger after each expansion stage. The same correlations as those used for modeling the intercoolers could be used here as well. One needs to just consider the physical properties of the secondary working fluid of these heat exchangers, which is surely much different than the previous heat exchangers. For modeling the heat exchangers after the turbines, the same model as the charging phase heat exchangers can be used. In fact, in case of using the heat and cold production of an SCAES unit for district energy systems, for an intercooler, the secondary working fluid will be warm pressurized water. For these heat exchangers, the inlet secondary fluid is the return line of the district cooling system with a temperature of about 15°C, while its flow rate is controlled to get an exit temperature of 8°C [37]. Then, the amount of cold energy collected from these heat exchangers can be calculated by:

$$\dot{Q}_c = \sum_{j=1}^{n} \left(\dot{m}_f \left(T_{f,i} - T_{f,e} \right) \right)_j \tag{7.10}$$

In the end, in the charging operation, the air storage reservoir pressure will be calculated by:

$$P_{asr}^{\lambda} = \left(\frac{m_{asr} R T_{asr}}{V_{asr}} \right)^{\lambda} \text{ where} : m_{asr}^{\lambda} = m_{asr}^{\lambda-1} - \dot{m}_a \tag{7.11}$$

Finally, having the model of the SCAES system and knowing that the SCAES unit will tri-generate cold, heat, and power over a roundtrip charging-discharging operation, the following terms of energy efficiency are defined for the performance of the system:

$$\eta_h = \sum_{\lambda=1}^{tc} \left(\frac{\dot{Q}_h}{\dot{E}_s} \right)^{\lambda} \tag{7.12}$$

$$\eta_c = \frac{\sum_{\lambda=1}^{td} \dot{Q}_c^{\lambda}}{\sum_{\lambda=1}^{tc} \dot{E}_s^{\lambda}} \tag{7.13}$$

$$\eta_p = \frac{\sum_{\lambda=1}^{td} \dot{E}_d^{\lambda}}{\sum_{\lambda=1}^{tc} \dot{E}_s^{\lambda}} \tag{7.14}$$

$$COP = \frac{\sum_{\lambda=1}^{td} \dot{E}_d^{\lambda} + \sum_{\lambda=1}^{td} \dot{Q}_c^{\lambda} + \sum_{\lambda=1}^{tc} \dot{Q}_h^{\lambda}}{\sum_{\lambda=1}^{tc} \dot{E}_s^{\lambda}} = \eta_{pth} + \eta_{ptp} + \eta_{ptc} \tag{7.15}$$

In these correlations, the first one defines the heat production efficiency, the second correlation is the cold efficiency, the third is the electricity generation efficiency, and the last is the overall energy efficiency of the SCAES system. Note that the parameters tc and td are the numbers of time-steps of the system operating in the charging and discharging phases, respectively.

7.2.2 HTHPS technology

As discussed, HTHPS systems come in two different designs. Thus, the mathematical model of each is presented separately. However, as both of these systems have the same element as the energy source supplier (i.e., a packed bed of rocks that is heated up by an airflow first passing over hot electrical coils), the model associated with the packed bed of stones is given here first.

There are a variety of models for calculating the variation of the packed bed storage temperature over the charging and discharging processes. One of the most reliable models is the one-dimensional, two-phase Schumann model [38], based on which the heat transfer through the stone particles is considered perfectly uniform towards the radius of the storage. Therefore, the Schumann model considers a non–uniform heat distribution along the height, in the same direction as the axial airflow (from top to bottom in the charging mode and vice versa for the discharging phase), given by the following set of correlations:

$$\rho_s c_s (1 - \epsilon) \frac{dT_s}{dt} = h_v (T_a - T_s) + \overline{k} \frac{d^2 T_s}{dx^2} \tag{7.16}$$

$$\rho_a c_{p,a} \epsilon \frac{dT_a}{dt} = h_v (T_s - T_a) - \frac{\dot{m}_a c_{p,a}}{A} \frac{dT_a}{dx} - \frac{D\pi U}{A} (T_a - T_{soil}) \tag{7.17}$$

$$h_v = 700 \left[\frac{\dot{m}_a}{Ad} \right]^{0.76} \tag{7.18}$$

$$\overline{k} = k_s (1 - \epsilon) + k_a \epsilon \tag{7.19}$$

$$U = \left(\frac{1}{\overline{h}_{in}} + \frac{R}{k_{ins}} \ln \left(\frac{R + \delta_{ins}}{R} \right) + \frac{R}{k_s} \ln \left(\frac{R + \delta_{ins} + \overline{R}}{R + \delta_{ins}} \right) \right)^{-1} \tag{7.20}$$

In these equations, ρ is density, c_p is thermal capacity at constant pressure, T is temperature, $\dot{m}$ is the mass flow rate, ϵ is the porosity factor of the packed bed, A is the cross-section area, D is the diameter, h_v is the volumetric heat transfer coefficient, $\bar{k}$ is the effective conductivity, and U is the overall heat transfer coefficient from the storage to the surrounding. Also, the parameters d, k, $\bar{h}$, δ, R, and $\bar{R}$ are the equivalent diameter, the conductivity coefficient, the convective heat transfer coefficient, the insulation thickness, the radius, and the thermal influential distance of the storage. The subscripts a, s, *soil*, *in*, and *ins*, respectively, refer to the airflow, the stone particles, the soil around the packed bed, the inner area of the packed bed, and its insulation.

The Schumann model considers all the rock particles in the storage as uniform particles with Biot numbers less than 0.1 to be able to take the stone particles as lumped bodies. In addition, the model neglects the effects of changes in the pressure and viscosity of the air and stones on the heat transfer process within the storage [39].

Having modeled the charging/discharging process of a packed bed of stones, which is similar for both air-based and steam-based HTHPS systems, the models of the power blocks of these two concepts are individually presented in the sections that follow.

7.2.2.1 Air-based

As mentioned, the discharging operation of an air-based HTHPS consists of parallel operation of the packed bed of stones with a power block, which is a multi-stage Bryton cycle with intercoolers and preheaters. Therefore, the first component of the system to be modeled is the compression stage including the compressor stages and the intercoolers. For this, the same correlations as those used for the SCAES system (i.e., Eqs. 7.1–7.6) can be used. The only difference here is that the compressors of the power block of the air-based HTHPS system operate in series all the time, while for an SCAES system, the arrangement of the compressor stages changes according to the pressure of the air storage reservoir.

The next component is the expansion part including the heat exchangers supplying the required heat of the airflow to reach the desired temperature before expanders and the expander stages. For the preheating heat exchangers the same correlations as those used for intercoolers can be used. The only difference here is that the preheaters are not gas-fluid heat exchangers, which present a high effectiveness factor, but rather they are air-air heat exchangers for which it is quite difficult to achieve high effectiveness. For gas-gas heat exchangers, an effectiveness factor up to 0.85 may be achievable.

For the expanders, we can use the same correlation as that used for the air expanders of the SCAES system (Eqs. 7.8–7.9). Here, the isentropic efficiency of the turbines could be much greater (even up to 90%) as the air is heated before the expansion and thus it will reach too-low temperatures after the expansion process. Rather, the design of the system is such that the exit air temperature of the expanders is quite high after each expansion stage so that it can be taken back into the packed bed to slow down the discharging pace of the storage.

Finally, having the energy conversion efficiency of the electricity generator as well as the work of the compressor and turbine sets calculated, the power production rate of the air-based HTHPS storage system is given as:

$$\dot{E}_{ab-hthps} = \eta_g \left(\dot{W}_t - \dot{W}_c \right) \tag{7.21}$$

As the system co-generates electricity and heat as discharging, for evaluating the energy performance of the system, the electricity-to-electricity and electricity-to-heat efficiencies of the system are defined as:

$$\eta_{p-ab-hthps} = \left. \sum_{t=1}^{td} \dot{E}_{ab-hthps} \middle/ \sum_{t=1}^{tc} \dot{E}_s \right. \tag{7.22}$$

$$\eta_{h-ab-hthps} = \left. \sum_{t=1}^{td} \sum_{i=1}^{n} \dot{Q}_h \middle/ \sum_{t=1}^{tc} \dot{E}_s \right. \tag{7.23}$$

In the last equation, n refers to the number of heat exchangers that supply heat to district heating (or other heating applications).

7.2.2.2 Steam-based

For a steam-based HTHPS, the power block is a Rankine cycle. For doing an energy analysis of a Rankine cycle, the boiler is the first component to be modeled. For a boiler with two regeneration loops as shown in Fig. 7.4, the heating duty can be calculated as:

$$\dot{Q}_{blr} = \dot{m}_{st}(h_e - h_i)_{st} + \dot{m}_{rg-1}(h_e - h_i)_{rg-1} + \dot{m}_{rg-2}(h_e - h_i)_{rg-2} \tag{7.24}$$

In which *blr* stands for the boiler, *st* refers to the main steam/water flow that is superheated through the boiler, and *rg-1* and *rg-2* are the first and second regeneration loops. The heating duty of the boiler ($\dot{Q}_{blr}$) is the rate of heat withdrawal from the packed bed of stones.

The next component is the turbines set, for which, assuming a triple-stage turbine as shown in Fig. 7.4, one could write:

$$\dot{W}_t = \dot{W}_{hpt} + \dot{W}_{ipt} + \dot{W}_{lpt}$$
$$= \dot{m}_{hpt}(h_i - h_e)_{hpt} + \dot{m}_{ipt}(h_i - h_e)_{ipt} + \dot{m}_{lpt}(h_i - h_e)_{hpt} \qquad (7.25)$$

In which the subscripts *hpt, ipt,* and *lpt* stand for the high-pressure, medium-pressure, and low-pressure turbines, respectively. The mass flow rate through these turbines will be the same if there is no preheating process line in the cycle. In many cases, a portion of the steam is withdrawn for preheating the water flow before entering the boiler to increase the efficiency of the cycle. The summation of the work produced by all the turbine stages gives the total work production of the cycle.

Calculating the work of the pumps of the system, one finds the net work production of the cycle and thereby the electricity generation rate of the cycle. These two parameters are given by:

$$\dot{W}_p = \dot{m}_p(h_i - h_e)_p \Big/ \eta_{is,p} \qquad (7.26)$$
$$\dot{E}_{sb-hthps} = \eta_g \left(\dot{W}_t - \dot{W}_p \right)$$

Where the term $\eta_{is,\,p}$ refers to the isentropic efficiency of the pump.

The following correlation also gives the rate of heat rejected from the steam through the condenser, which is equal to the supplied heat to the district heating system (or used for other heating purposes).

$$\dot{Q}_{cond} = \dot{m}_{cond}(h_i - h_e)_{cond} = \dot{Q}_h \qquad (7.27)$$

Finally, the electricity and heat generation efficiencies of the steam-based HTHPS system are calculated, respectively, by:

$$\eta_{p-sb-hthps} = \sum_{t=1}^{td} \dot{E}_{sb-hthps} \Big/ \sum_{t=1}^{tc} \dot{E}_s \qquad (7.28)$$

$$\eta_{h-sb-hthps} = \sum_{t=1}^{td} \dot{Q}_{cond} \Bigg/ \sum_{t=1}^{tc} \dot{E}_s \qquad (7.29)$$

The overall efficiency of the system equals the summation of the above two efficiency factors.

7.2.3 GES technology

This section of the chapter gives the mathematical model of a ground-GES system only. For the other configurations of GES technology, based on the same fundamentals, the mathematical model could be developed.

In a ground-GES system as shown in Fig. 7.5, the charging process is just pumping water to high pressures to give a higher elevation to the rock piston. In this system, the amount of energy to be stored is mainly due to the potential energy made by elevating the piston and partially by the water height below the piston as it goes higher. Therefore, the amount of energy to be stored in this system is calculated as:

$$E_{stored} = \int_0^H (Mg + \rho Agx)dx \qquad (7.30)$$

In which M is the mass of the rock piston, H is the height of the piston after the charging process (assuming a height of zero at the beginning of the charging), A is the cross-sectional area of the piston, ρ is water density, and x refers to the upward direction.

Considering the overall energy losses of the process (the electricity generator losses as well as mechanical losses), the amount of power to be produced as the piston is released can be calculated by:

$$E_{g-ges} = \eta_g \eta_{mec} E_{stored} \qquad (7.31)$$

The required work of the pump to dislocate the piston from a zero elevation point to the height of H can be given by the Bernoulli equation. Neglecting the losses through the water channel/piping, based on the Bernoulli equation for a steady, incompressible, and inviscid flow, the head of the fluid changing conditions/properties from state 1 to state 2 (which is associated with the losses that the pump should compensate to make the required change in the fluid conditions happen), is calculated by:

$$H = \left(\frac{P}{\rho g} + \overbrace{\frac{Ve^2}{2g}}^{KE} + \overbrace{z}^{PE} \right)_2 - \left(\frac{P}{\rho g} + \frac{Ve^2}{2g} + z \right)_1 \qquad (7.32)$$

In which Ve is the velocity of the fluid and z is its elevation; all the other symbols have already been introduced in the previous correlations. As indicated by the formula, the second and third terms in the parentheses are associated with the kinetic and potential energies. Naturally, the pressure of the fluid at the free surface position (the water source) is equal to the ambient pressure, and the pressure of the second point (which is just after the pumps) is extremely high due to the effect of the weight of the piston (and water below the piston). Therefore, the head could be estimated by neglecting the effect of kinetic energy and elevation changes. The pressure at point 2 is:

$$P_2 = \frac{F}{A} = \frac{(M + M_w)g}{\pi r^2} \qquad (7.33)$$

In which M is the mass of the piston, M_w is the mass of water below the piston in the hole, which requires integration over the movement course of the piston because it increases as the charging process continues, and r is the radius of its cross-section. F and A are the force of weights and the piston's cross-sectional area.

The power to be received by the fluid to reach this head is calculated by:

$$W_H = \rho g Q H \qquad (7.34)$$

Where Q is the quantity of water to be pumped. Finally, the following equation gives the work of the pump:

$$W_P = \frac{W_H}{\eta_{mec,p}} = \frac{\rho g Q H}{\eta_{mec,p}} \qquad (7.35)$$

In which $\eta_{mec,p}$ is the mechanical efficiency of the pumps.

Finally, the efficiency of this system over a roundtrip charging-discharging process can be defined as:

$$\eta_{p-g-ges} = E_{g-ges} / W_p \qquad (7.36)$$

7.3 Future perspective

This chapter introduced and discussed three specific types of MES systems that seem more inspiring than other new emerging systems in this category. Each of these systems presents its own advantages and disadvantages, which affect their future perspectives in both positive and negative ways.

Regarding the SCAES system, the very inspiring point is that it offers trigeneration of cold, heat, and electricity. This is a very important option of SCAES as it can be used to integrate different energy sectors, which is one of the most important characteristics of future smart energy systems [40]. The overall efficiency of SCAES is also very impressive, making the system cost-effective. Another advantage is the fairly acceptable performance of the system in even very low load levels compared to nominal operating conditions. Although that is mainly the heat output of the system, and both cold and electricity outputs approach zero as the load level drops significantly, the high heat output rate of the system in such operating conditions, which are inevitable in many energy storage units, keeps the system cost-effective in real-life operating conditions [20]. Finally, as was previously explained, the agility of energy storage systems is especially important, particularly in the electricity sector where SCAES technology is highly agile compared to other CAES technologies due to the fast expansion process in which no heating process is required for the compressed air. On the other hand, the SCAES system, like other CAES designs, suffers from geographical limitations. It requires digging an appropriate salt or rock cavern, unless the system is designed for use on smaller scales for which aboveground storage vessels could be used. The other disadvantage of the technology is its low electricity efficiency, although its overall efficiency is outstanding. This is of importance because the electricity grid is the dominant energy sector in any energy systems at the moment (including those energy systems with distributed cold and heat networks) and this will be even more important in the future because of two main reasons. The first is the well-developed state of practice of renewable-based power generation technologies, such as wind turbines and PV panels [41]. Second is the increasing penetration of electrical-driven facilities in the other energy sectors (e.g., electrical vehicles [42], large-scale heat pumps for district cooling and heating [13], etc.). Therefore, efforts for improving the design of this system to present greater electricity efficiency are vital for making the future applicability of this technology promising. This is why Alsagri et al. [21] proposed the combination of an SCAES system with an organic Rankine cycle in a way that the heat generated during the charging mode of the system can be used to drive the Rankine cycle. Then, the generated power by the organic cycle is used to drive the compressors. In this way, the required size of the compression unit decreases and a significant improvement in the electricity generation efficiency of the SCAES system is achieved. Improving the performance of the cold air expanders is a way to achieve better electricity efficiency of the cycle. The expander type mainly being used in studies on this system includes single-screw expanders with a maximum isentropic efficiency of 65% [43]. For the real-life implementation of this technology, a pilot-scale demonstration is a prerequisite. There is hope that this will happen in either Denmark or China in the near future, as studies on this system mainly originated from these two countries.

With respect to HTHPS systems, regardless of category, the same problems as those of SCAES technology are seen, that is, the fairly low electricity efficiencies that these systems offer. Therefore, no matter what advantages or disadvantage these systems may offer, one needs to focus on improving the electrical efficiencies of these two innovative systems. Arabkoohsar [25, 44] has proposed and assessed the hybridization of these two systems with a variety of organic Rankine units in the discharging mode to increase the share of electricity generation of the system compared to its heat generation. Both of the investigations show great potential to significantly improve this feature of HTHPS systems. Steam-based will not be able to significantly contribute to heat supply after combination with an organic cycle, but the combined air-based design still generates much heat for district heating supply. On the other hand, both HTHPS designs offer co-generation of heat and electricity, which is appropriate for future integrated smart energy systems or even existing energy systems in which power and heat grids are well distributed. The other important common advantage of these systems is that the medium of storing heat is cheap and not limited by any geographical or special needs. This makes the technology cost-effective and implementable everywhere. A Rankine-based HTHPS system is very slow in starting up, while an energy storage unit must be fast enough in changing the load, going to stand-by mode, and coming into operation. Therefore, maybe the use of this system as a pure energy storage unit does not make sense. Rather, this system could be used as a power plant that is connected to a large-scale packed bed of stones for storing the electricity of the plant itself when the electricity grid spot price is low or for storing the electricity of a renewable power plant in the network when surplus power is generated. Then, when the electricity price goes high or the renewable plant needs compensation for the ramps, the stored heat can be reclaimed for boosting Rankine cycle production. The air-based system is, on the other hand, very good in this aspect, like a gas turbine that can come into operation pretty quickly. Compared to the SCAES system, however, the air-based system is less agile as real-time compression of air and then preheating that air is required before the expansion process. None of this is needed in an SCAES system. Overall, the technology shows an inspiring perspective for coming into service as a component of real energy systems. As mentioned, Germany has started using this technology in steam-based design, but there is much hope that the air-based system also gets the attention of the relevant role makers in the market as it is more advantageous than the steam-based system as an energy storage unit in many aspects.

Talking about the GES technology, it depends on which category you are talking about. A ground-GES is quite interesting as it is not restricted by special geographical needs. For ground-GES, a large and deep hole is needed, which is most likely implementable everywhere. The piston of this system is proposed to be of rock material that is readily available at very low cost. Rocks also offer good density, making the energy density of the system acceptable. Aluminum, copper, iron, and others are alternative materials to rocks. Although they offer greater densities, they will probably not result in positive economic outcomes of the system. Ref. [33] recommends iron as the most suitable material for the piston. The efficiency of the system is also expected to be very high in the range of 80%–90%. As the main force source of the system is the weight of the piston, it does not go to low operational load and thus neither the pump and motor nor the turbine and generator are seriously affected by partial load operating conditions. A serious drawback of this technology is, however, the lack of applied research and development on it. To make this system a serious option as compared to other MES systems, development and demonstration at the laboratory and then pilot scale is needed. Regarding ocean-GES, unlike ground-GES for which simplicity is a key advantage, the ocean-GES system is a bit complicated. Having submarine robots connect and disconnect the cables to several objectives (weights) underwater is, for example, one of the complexities of this system. The fact that the system needs very deep water to offer better storage capacity makes the system less cost-effective because this means a quite large distance from the consumption/production locations. Deepwater is usually far away from lands (islands). This technology also does not have a rich literature. To make it an option for storage applications, more demonstration studies are needed.

References

[1] A. Arabkoohsar, L. Machado, M. Farzaneh-Gord, R.N.N. Koury, Thermo-economic analysis and sizing of a PV plant equipped with a compressed air energy storage system. Renew. Energy 83 (2015), https://doi.org/10.1016/j.renene.2015.05.005.

[2] B. Elmegaard, W. Brix, Efficiency of compressed air energy storage, in: Proceedings of the 24th International Conference on Efficiency, Cost, Optimization, Simulation and Environmental Impact of Energy Systems ECOS, vol. 2011, 2011, pp. 2512–2523.

[3] H. Peng, Y. Yang, R. Li, X. Ling, Thermodynamic analysis of an improved adiabatic compressed air energy storage system. Appl. Energy 183 (2016) 1361–1373, https://doi.org/10.1016/j.apenergy.2016.09.102.

[4] M. Heidari, S. Wasterlain, P. Barrade, F. Gallaire, A. Rufer, Energetic macroscopic representation of a linear reciprocating compressor model. Int. J. Refrig. 52 (2015) 83–92, https://doi.org/10.1016/j.ijrefrig.2014.12.019.

[5] A. Arabkoohsar, L. Machado, R.N.N. Koury, K.A.R. Ismail, Energy consumption minimization in an innovative hybrid power production station by employing PV and evacuated tube collector solar thermal systems. Renew. Energy 93 (2016) 424–441, https://doi.org/10.1016/j.renene.2016.03.003.

[6] A. Arabkoohsar, L. Machado, R.N.N. Koury, Operation analysis of a photovoltaic plant integrated with a compressed air energy storage system and a city gate station. Energy 98 (2016) 78–91, https://doi.org/10.1016/j.energy.2016.01.023.

[7] A. Arabkoohsar, L. Machado, M. Farzaneh-Gord, R.N.N. Koury, The first and second law analysis of a grid connected photovoltaic plant equipped with a compressed air energy storage unit. Energy 87 (2015) 520–539, https://doi.org/10.1016/j.energy.2015.05.008.

[8] A. Odukomaiya, A. Abu-Heiba, K.R. Gluesenkamp, O. Abdelaziz, R.K. Jackson, C. Daniel, S. Graham, A.M. Momen, Thermal analysis of near-isothermal compressed gas energy storage system. Appl. Energy 179 (2016) 948–960, https://doi.org/10.1016/j.apenergy.2016.07.059.

[9] S. Briola, P. Di Marco, R. Gabbrielli, J. Riccardi, A novel mathematical model for the performance assessment of diabatic compressed air energy storage systems including the turbomachinery characteristic curves. Appl. Energy 178 (2016) 758–772, https://doi.org/10.1016/j.apenergy.2016.06.091.

[10] E. Barbour, D. Mignard, Y. Ding, Y. Li, Adiabatic compressed air energy storage with packed bed thermal energy storage. Appl. Energy 155 (2015) 804–815, https://doi.org/10.1016/j.apenergy.2015.06.019.

[11] D. Wolf, M. Budt, LTA-CAES—a low-temperature approach to adiabatic compressed air energy storage. Appl. Energy 125 (2014) 158–164, https://doi.org/10.1016/j.apenergy.2014.03.013.

[12] A. Arabkoohsar, M. Dremark-Larsen, R. Lorentzen, G.B. Andresen, Subcooled compressed air energy storage system for coproduction of heat, cooling and electricity. Appl. Energy 205 (2017) 602–614, https://doi.org/10.1016/j.apenergy.2017.08.006.

[13] A. Arabkoohsar, Non-uniform temperature district heating system with decentralized heat pumps and standalone storage tanks. Energy 170 (2019) 931–941, https://doi.org/10.1016/j.energy.2018.12.209.

[14] W. Sun, S. Chen, Y. Hou, S. Bu, Z. Ma, L. Pan, Numerical studies on the off-design performance of a cryogenic two-phase turbo-expander. Appl. Therm. Eng. 140 (2018) 34–42, https://doi.org/10.1016/j.applthermaleng.2018.05.047.

[15] A. Arabkoohsar, G.B. Andresen, Design and optimization of a novel system for trigeneration. Energy 168 (2019) 247–260, https://doi.org/10.1016/j.energy.2018.11.086.

[16] A. Arabkoohsar, G.B. Andresen, Supporting district heating and cooling networks with a bifunctional solar assisted absorption chiller. Energy Convers. Manag. 148 (2017) 184–196, https://doi.org/10.1016/j.enconman.2017.06.004.

[17] A. Arabkoohsar, An integrated subcooled-CAES and absorption chiller system for cogeneration of cold and power, in: IEEE Xplore, Proceeding SEST 2018, 2018, pp. 1–5.

[18] M. Sadi, A. Arabkoohsar, Modelling and analysis of a hybrid solar concentrating-waste incineration power plant. J. Clean. Prod. (2018), https://doi.org/10.1016/j.jclepro.2018.12.055.

[19] B.V. Mathiesen, H. Lund, D. Connolly, H. Wenzel, P.A. Østergaard, B. Möller, S. Nielsen, I. Ridjan, P. Karnøe, K. Sperling, F.K. Hvelplund, Smart energy systems for coherent 100% renewable energy and transport solutions. Appl. Energy 145 (2015) 139–154, https://doi.org/10.1016/j.apenergy.2015.01.075.

[20] A.S. Alsagri, A. Arabkoohsar, H.R. Rahbari, A.A. Alrobaian, Partial load operation analysis of Trigeneration subcooled compressed air energy storage system. J. Clean. Prod. (2019) 117948, https://doi.org/10.1016/j.jclepro.2019.117948.

[21] A.S. Alsagri, A. Arabkoohsar, Combination of subcooled compressed air energy storage system with an organic Rankine cycle for better electricity efficiency, a thermodynamic analysis, J. Clean. Prod. 239 (2020).

[22] A. Arabkoohsar, G.B. Andresen, Design and analysis of the novel concept of high temperature heat and power storage. Energy 126 (2017) 21–33, https://doi.org/10.1016/j.energy.2017.03.001.

[23] Stiesdal, Stiesdal Electricity Storage, https://www.stiesdal.com/energy-storage/, 2020.

[24] T. Esence, A. Bruch, S. Molina, B. Stutz, J.-F. Fourmigué, A review on experience feedback and numerical modeling of packed-bed thermal energy storage systems. Sol. Energy 153 (2017) 628–654, https://doi.org/10.1016/j.solener.2017.03.032.

[25] A. Arabkoohsar, Combined steam based high-temperature heat and power storage with an organic Rankine cycle, an efficient mechanical electricity storage technology. J. Clean. Prod. (2019), https://doi.org/10.1016/j.jclepro.2019.119098.

[26] A. Arabkoohsar, G.B. Andresen, Thermodynamics and economic performance comparison of three high-temperature hot rock cavern based energy storage concepts. Energy 132 (2017), https://doi.org/10.1016/j.energy.2017.05.071.

[27] A. Arabkoohsar, G.B. Andresen, Dynamic energy, exergy and market modeling of a high temperature heat and power storage system. Energy 126 (2017), https://doi.org/10.1016/j.energy.2017.03.065.

[28] A.S. Alsagri, A. Arabkoohsar, M. Khosravi, Efficient and cost-effective district heating system with decentralizedheat storage units, and triple-pipes, Energy 188 (2019) 116035.

[29] H. Nami, A. Arabkoohsar, Improving the power share of waste-driven CHP plants via parallelization with a small-scale Rankine cycle, a thermodynamic analysis. Energy 171 (2019) 27–36, https://doi.org/10.1016/j.energy.2018.12.168.

[30] A. Arabkoohsar, A.S. Alsagri, A new generation of district heating system with neighborhood-scale heat pumps and advanced pipes, a solution for future renewable-based energy systems. Energy 193 (2020), https://doi.org/10.1016/j.energy.2019.116781.

[31] Siemens, Siemens High Temeprature Heat and Power Storage Project, https://www.siemens.com/press/en/pressrelease/?press=/en/pressrelease/2016/windpower-renewables/pr2016090419wpen.htm&content=WP, 2016.

[32] T. Morstyn, M. Chilcott, M.D. McCulloch, Gravity energy storage with suspended weights for abandoned mine shafts. Appl. Energy 239 (2019) 201–206, https://doi.org/10.1016/j.apenergy.2019.01.226.

[33] C.D. Botha, M.J. Kamper, Capability study of dry gravity energy storage. J. Energy Storage 23 (2019) 159–174, https://doi.org/10.1016/j.est.2019.03.015.

[34] A.S. Alsagri, A.D. Chiasson, Thermodynamic analysis and multi-objective optimizations of a combined recompression SCO2 Brayton Cycle-TCO 2 Rankine cycles for waste heat recovery, Int. J. Curr. Eng. Technol. 8 (2018) 541–548.

[35] H. Guo, Y. Xu, Y. Zhang, Q. Liang, H. Tang, X. Zhang, Z. Zuo, H. Chen, Off-design performance and an optimal operation strategy for the multistage compression process in adiabatic compressed air energy storage systems. Appl. Therm. Eng. 149 (2019) 262–274, https://doi.org/10.1016/j.applthermaleng.2018.12.035.

[36] W. He, Y. Wu, Y. Peng, Y. Zhang, C. Ma, G. Ma, Influence of intake pressure on the performance of single screw expander working with compressed air. Appl. Therm. Eng. 51 (2013) 662–669, https://doi.org/10.1016/j.applthermaleng.2012.10.013.

[37] A. Arabkoohsar, G.B. Andresen, A smart combination of a solar assisted absorption chiller and a power productive gas expansion unit for cogeneration of power and cooling. Renew. Energy 115 (2018) 489–500, https://doi.org/10.1016/j.renene.2017.08.069.

[38] K. Attonaty, P. Stouffs, J. Pouvreau, J. Oriol, A. Deydier, Thermodynamic analysis of a 200 MWh electricity storage system based on high temperature thermal energy storage. Energy 172 (2019) 1132–1143, https://doi.org/10.1016/j.energy.2019.01.153.

[39] A. Benato, Performance and cost evaluation of an innovative pumped thermal electricity storage power system. Energy 138 (2017) 419–436, https://doi.org/10.1016/j.energy.2017.07.066.

[40] H. Lund, N. Duic, P.A. Østergaard, B.V. Mathiesen, Smart energy systems and 4th generation district heating. Energy 110 (2016) 1–4, https://doi.org/10.1016/j.energy.2016.07.105.

[41] M. Diesendorf, B. Elliston, The feasibility of 100% renewable electricity systems: a response to critics. Renew. Sust. Energ. Rev. 93 (2018) 318–330, https://doi.org/10.1016/j.rser.2018.05.042.

[42] J. Kester, L. Noel, G. Zarazua de Rubens, B.K. Sovacool, Policy mechanisms to accelerate electric vehicle adoption: a qualitative review from the Nordic region. Renew. Sust. Energ. Rev. 94 (2018) 719–731, https://doi.org/10.1016/j.rser.2018.05.067.

[43] W. Wang, Y.T. Wu, G.D. Xia, C.F. Ma, J.F. Wang, Y. Zhang, Experimental study on the performance of the single screw expander prototype by optimizing configuration. in: ASME 2012 6th International Conference on Energy Sustainability. ES 2012, Collocated With ASME 2012 10th International Conference on Fuel Cell Science, Engineering and Technology, American Society of Mechanical Engineers, 2012, pp. 1281–1286, https://doi.org/10.1115/ES2012-91502.

[44] A. Arabkoohsar, Combination of air-based high-temperature heat and power storage system with an organic Rankine cycle for an improved electricity efficiency. Appl. Therm. Eng. 167 (2020) 114762, https://doi.org/10.1016/j.applthermaleng.2019.114762.

CHAPTER EIGHT

Conclusion

Ahmad Arabkoohsar
Department of Energy Technology, Aalborg University, Esbjerg, Denmark

Abstract

Today wind turbine technology has reached a point that its electricity has become cheaper than conventional energy. This is even true for locations with wind capacity factors less than 20%. However, the large-scale development of renewable energy technologies, including wind, is limited because they are intermittent. For example, for a wind turbine, if there is less wind, there will be less or even no power output. The same applies for solar-driven power plants. For making controllable frequency grids, the fluctuations of solar and wind plants must be managed. For example, agile conventional technologies such as gas turbines can come into operation or go for less production to compensate for sudden up-and-down changes in the power output of renewable power plants. The objective is to minimize and even eliminate the contribution of conventional plants as they seriously contribute to global warming. Therefore, storing the energy of renewable power plants when over-producing, and reclaiming it back to the grid when needed such as during peak consumption periods or low-production times, is vital for energy systems with high penetration of solar and wind plants. In addition, such electricity storage units give the opportunity to levelize the cost of production and electricity market spot pricing by storing electricity when the spot price is low and giving it back to the market when the spot price gets too high. This book discussed the limitations of the widely used electricity storage solutions and introduced an alternative category of solutions, mechanical energy storage (MES) technologies. Chapter 2 discussed thermal energy storage (TES) systems, Chapter 3 examined compressed air energy storage (CAES), Chapter 4 covered pumped hydropower storage (PHS), Chapter 5 discussed flywheel energy storage (FES), Chapter 6 focused on pumped heat electricity storage (PHES), and Chapter 7 examined three newly introduced systems of subcooled compressed air energy storage (SCAES), trigeneration compressed air energy storage (TCAES), and high-temperature heat and power storage (HTHPS). In this chapter, we present our final remarks and conclusions.

8.1 Discussion of mechanical energy storage (MES) systems

8.1.1 Thermal energy storage (TES) systems

Sensible, and to a lesser extent, latent thermal energy storage (TES) techniques have received more attention in the literature than chemical and thermochemical thermal storage techniques [1]. Seasonal TES systems, TES units of solar thermal plants, and distributed TES units for domestic uses are among the most broadly available systems [2]. They are classified into large-, medium-, and small-scale systems as well as into high-, medium-, and low-temperature systems [3].

Low-temperature, small-scale heat storage units are appropriate for domestic heat storage uses, whereas low-temperature, large-scale units are used for seasonal storage systems [4]. Except for the low-temperature small- and even large-scale TESs which cannot be relevant to MES systems for electrical storage applications, the literature shows the possible application of the other categories of TESs for this objective in different ways. Medium- and large-scale, medium-temperature systems can be used as heat storage units for different processes such as heat collection and supply of low-temperature compressed air energy storage (CAES) systems [5] or the medium-temperature solar thermal systems used to cover part of the heat demand of such systems [6], and so forth. Medium- and large-scale high-temperature systems can be used for direct contribution to the process of electricity storage and production. Examples of this include the high-temperature heat storage unit of solar concentrating power plants [7] or the high-temperature packed beds for use in high-temperature heat and power storage (HTHPS) technology [8,9]. Most TES solutions for medium- and high-temperature heat storages (100–250°C and >250°C, respectively) are sensible heat storage materials such as pressurized water, industrial oils, rocks, metals, and so on. [10]. One of the challenges of sensible heat storage systems is the large volume required, as all the stored heat is supposed to be used for increasing the temperature of the storage medium. Therefore, as the possible temperature difference of the material is limited, the mass of the material of storage should increase. Before deciding about the use of such TES systems, some critical parameters such as local geological conditions, required size and temperature, legal aspects regarding drilling, for example, and investment costs should be taken into account. At high storage temperatures, there is a need for more care about the insulation, otherwise the stored heat will simply be dissipated to the surroundings [11]. Phase change materials (PCMs) (latent heat storage) such as molten salt are also commonly used for specific types of power storage such as solar power tower plants [12]. Most of these change phase from solid to liquid and vice versa. A critical problem of PCMs is their low heat conduction coefficient in the solid phase. The convective heat transfer when in liquid phase compensates for this deficiency. However, when in solid phase, the charging/discharging rates can be seriously affected. This has given the conductivity enhancement of PCMs a special importance for research [13].

It should also be noted that cold storage systems, which are still in the category of TES, will also be needed for power storage applications after the presentation of the two concepts of pumped heat electricity storage (PHES) [14] and Stiesdal StorageTechnology (SST), which is, in reality, a similar concept to PHES [15]. In these systems, there is a high-temperature heat storage unit (about 600°C) and one cold storage unit. The temperature of the cold storage unit of these systems is very low (−100°C). The literature states that some of the regular materials of TES can cover the minimum temperature of −40°C, while specific cold storage materials are required for temperatures as low as −100°C. The SST offers a layered packed bed of rocks for this [15].

Apart from the type and category of TES techniques just discussed, one could say that a high CAPEX requirement is a challenge for most of these techniques. Therefore, more research into reducing the costs of these or introducing cheaper yet stable storage materials for such severe heat/cold storage conditions is still required. Natural materials such as rocks have recently been introduced as TES solutions that well address this challenge though. Overall, for any new TES materials/techniques, the critical required features are low cost, high stability over a long lifetime, high storage density, fast charging/discharging rates, and low rate of losses.

8.1.2 Pumped hyrdropower storage (PHS) systems

PHS is the most mature MES technology with an efficiency of about 85%, which is the highest electricity-to-electricity efficiency among all MES systems. This high capacity can be a blessing for highly renewable-based energy systems, especially from a long-term perspective [16]. The agility of PHS is one of its most important advantages so that it can be used even as an ancillary service of electricity grids for frequency regulation, for example [17]. It could be said that in real-life application, PHS is the only commercially proven large-scale MES that is appropriate for capacities in the range of 100 MW to a couple of thousand MWs. Long life span and long discharge time are further advantages of PHS technology [18]. Today, there is more than 125 GW running PHS capacity all around the world, which is equal to 3% of the whole electricity production of the world and 99% of the world's electricity storage capacity [19].

In spite of a relatively low production cost, PHS is almost as expensive as a battery. There is not a good chance to reduce the costs associated with energy conversion system equipment due to its mature state of practice. Geographical restrictions, direct and indirect negative environmental impacts, too-long lead construction time, and the very large area needed are other important challenges of PHS technology. The environmental problems are associated with the inevitable changes in nature for the required large dams, roads, and power lines. A serious technical challenge of PHS is sediment management. During the lifetime of the PHS system (and regular hydropower systems), sediment is gradually trapped in the dam and the infrastructure causes a lot of environmental, operational, and maintenance difficulties such as decreasing the production capacity of the system, damaging the turbines, and so on [20]. A very thorough discussion of trends and challenges of PHS can be found in Ref. [21].

Hydropower including PHS is and will continuously play a key role in the global energy matrix. However, due to the seriousness of the aforementioned challenges, there have always been efforts to raise new solutions in this context. Some of these are PHS reservoirs as wastewater treatment storage, underground water-filled shafts with piston-floated mechanisms, undersea PHS for offshore wind turbines, underground PHS, and others [22]. Even with these novel concepts, there is still a need for more innovations and practical solutions, including new technological and economic possibilities, to appropriately address the critical challenges of PHS systems.

8.1.3 Flywheel energy storage (FES) systems

An FES system offers several advantages making them an appropriate energy storage technology for short- or medium-term applications; up to up to 133 kWh in some advanced designs [23]. A FES system is environmentally friendly, does not suffer from any serious depreciation over numerous charge/discharge cycles, requires little maintenance, presents a long lifetime, and offers a high energy efficiency of up to 90% [24]. On the other hand, the main drawbacks of FES systems include high cost and very low energy density, which limits its application to small-scale and very specific needs, such as space applications [25].

Therefore, the critical focus for the future generation of FES technologies must be on reducing its costs and finding ways to increase its energy storage density. It is also found out that there might be some ways to reduce the mechanical, electrical, and energy conversion losses of FES systems and thereby obtain even greater efficiency from them. To address these issues, one of the focuses of research in this area has been on finding ways to increase the speed of rotation using multilevel converters and advanced technologies for manufacturing semiconductors. These will lead to an increase in the power density of FES systems. It is important to note that to increase the speed of FES systems, using ultrahigh-speed machines (up to 1000,000 rpm) is under investigation. The speed of existing systems is limited to almost one-tenth of this. One more research focus is on better materials to improve the strength of highly stressed rotors (especially when the speed is further increased) [26].

8.1.4 Compressed air energy storage (CAES) systems

CAES, in different configurations, is one of the storage technologies that has shown a promising future and possible widespread use. The levelized energy cost of CAES systems is expected to be much less than many of the other MES systems and way less than batteries [27]. This is mainly due to low capital cost, very long lifetime, high expected efficiency of up to 85%, flexibility in charging and discharging, and very large energy density [28]. One of the important advantages of CAES systems is the large possible capacities allowing for long charging and discharging periods of several hours or even days, which makes it an interesting solution for big energy systems [29]. The further key feature of the CAES system is its agility in coming into either charging or discharging operation, going to standby mode, or changing the operational load [30]. This and the long possible charging/discharging possibilities make the CAES capable of not only playing a key role in day-ahead and intraday electricity markets but also as an ancillary service in real-time markets [31]. The latter, of course, is more realistic for new generations of CAES such as trigeneration CAES [32] in which the expansion process is faster than other conventional CAES designs.

On the other hand, there is certainly a variety of issues that have limited the widespread deployment of CAES systems in real-life energy systems. One of the main issues is that CAES knowledge is still under development and has not reached a perfect mature state. A CAES system has a lower roundtrip efficiency than batteries and even some of the other MES systems such as PHS. A serious challenge in this context is that no off-the-shelf machinery appropriate for highly efficient CAES systems, especially advanced adiabatic CAES systems, is available yet. And, as already mentioned, geological restrictions and uncertainties are the most important issues with CAES systems.

To address these issues, several studies have been conducted and more are still required. These works focus (and should continue to focus) on finding solutions for speeding up even further the flexibility of CAES systems in terms of short start-up times and fast ramping. Subcooled CAES (SCAES) [27] and super high-pressure CAES combined with hydraulic components designed by Ref. [33] are the new designs of CAES that address this. Finding designs that do not require high-pressure, large-scale caverns to overcome the geological restrictions is the other focus of the literature. Ocean CAES is a solution for this, though it still requires access to deep seawater [34]. Finding better heat storage solutions with high energy densities [35], designing more efficient and less costly turbomachinery [36], and developing dynamic models for various designs of CAES systems to gain insight into their off-design performance and address the associated challenges [37–39] are some of the further frameworks of research activities on CAES systems.

One of the possibilities is to consider CAES systems not only for large-scale use, but rather for smaller scales for which the problem of special topology requirements is solved using simple, small, and high-pressure storage tanks [40]. At a small size, it will be much easier to overcome the existing economic barriers for CAES systems and thus will allow for much larger penetration of CAES systems (in different schemes) in a global scenario. This will be like using batteries in smart buildings where the buildings' own energy systems using the batteries can play a key role in peak shaving of the grid and offer almost net-zero electricity bills for the buildings over an entire year due to the interaction with the grid [41].

8.1.5 Pumped heat electricity storage (PHES) systems

PHES technology, also known as pumped thermal energy storage (PTES), is one of the concepts that has been under investigation in recent years. The main operating principle of a PHES system is storing electricity as a temperature difference between a very cold and a very hot thermal storage units between which a thermal engine is employed to take advantage of this temperature difference for power generation or storage [42]. The most well-known PHES designs include reversible Brayton cycle PHES (RBC-PHES), reversible transcritical organic Rankine cycle PHES (RTORC-PHES), and compressed heat energy storage PHES (CHES-PHES) systems, each of which has its own pros and cons. In an RBC-PHES system, there are two sensible thermal storage units (one cold and one hot) and a supercritical gas (usually air or argon) is used as the working fluid. In an RTORC-PHES system, supercritical CO_2 is used as the main working fluid and (one or more) ice storages are employed as the thermal storage units. Finally, a CHES-PHES system is based on a conventional steam Rankine cycle with a high-temperature PCM as the hot storage unit, while the cold storage is at ambient temperature [43].

In general, although it is a quite new technology, PHES systems are the most promising MES technologies because of their large energy density, long life cycle, freedom from geological restrictions, great roundtrip efficiency, acceptable cost, and capability to integrate with conventional power technologies. The energy density of a PHES system can be up to 170 kW per litter of thermal storage, the energy storage cost of that can be as low as 60 USD/kWh, and its roundtrip efficiency can reach up to 80% for about 30 years of useful lifetime [44]. This technology, however, still requires development. This includes appropriate turbomachinery technologies (especially, where the temperatures of the low and high operation levels get too severe through which the system efficiency can be greater), further investigation and optimization of the cold and hot storage materials and methods, and so forth [14].

8.1.6 New emerging technologies

It was discussed that several novel generations of MES systems are being continuously introduced to the literature. Among them are SCAES, HTHPS, and gravity energy storage (GES). Like the previously discussed MES systems, each of these offers some specific advantages but also suffers from some certain drawbacks.

The concept of SCAES is an inspiring scheme of CAES technologies that offers the tri-generation of heat, cold, and power at high overall efficiency [45]. In addition to the high coefficient of performance of about 1.5 (or even greater for advanced designs), the possibility of integrating cold, heat, and electricity sectors is an interesting feature. Fast response to the load changes and power ramps is a unique feature of this system. A possibility is developing small-scale SCAES systems for residential use to cover all the heat, electricity, and cold demand of the buildings. In this case, there is no need for an underground air storage reservoir either; simple high-pressure tanks are sufficient. This system, however, still suffers from geological restrictions for air storage reservoirs in case of large-scale use, very low electricity efficiency of about 30%, which is much less than other MES systems, and the special turbomachinery required for the very low operational condition of the turbines (around −100°C or so). Therefore, addressing all these challenges should be the main focus of future research works to push the technology further towards real-life deployment.

The concept of HTHPS is proposed in two designs: air-based and steam-based [46]. These are based on the idea of storing electricity as high-temperature thermal energy and then using it for co-generation of power and heat. The steam-based power block offers greater overall efficiency and greater electricity efficiency. However, it needs a higher CAPEX and, of course, it is not agile at all. Therefore, using it as an energy-only storage unit does not make sense. Rankine cycles take hours to come into useful operation in the case of shutdown [47], while an energy storage unit needs several running in and out of operation. However, using that as an accompanying part of an existing power plant like what Siemens A/S has recently done in Germany [48] for peak shaving or as an ancillary service is quite interesting. An air-based system, on the other hand, is based on a multi-stage gas turbine cycle. Thus, it is cheaper and more agile, but offers lower overall and electricity efficiency. This design of HTHPS can be a very good energy storage unit where both electricity and heat grids are available, making a reliable synergy between these two sectors. As the power blocks of these two systems are well known and quite well developed, there is not much chance that any advancements could be achieved in their current design; however, new designs or combinations of these systems with others for addressing their deficiencies are a possibility. The combination of these systems with organic Rankine cycles for improving their electrical efficiency are the two most recent works in this context [8, 9].

Finally, GES technology in several different schemes has been recently proposed and investigated. This technology may offer efficiencies up to 90% [49]. The utilization of gravity force either in seawater or on the ground are possibilities for this technology. Ground GES in any design is much more straightforward than an underwater GES system [50]. The former just needs a deep and large whole and a piston to rise up with either a cable or the force of water to be pumped below. The latter requires a quite complicated design and operation [51]. As these GES designs are quite new, there is not much information available for them in the literature. Therefore, extensive research on various aspects of these related to components, operation strategy, and so on is required to make the technology be considered a serious MES system among all the other possibilities.

8.2 Final remarks

There exist several energy storage technologies with well-developed and developing states of the art. Reviewing the advantages and disadvantages of any of these systems brought us to the conclusion that batteries as the most straightforward and widespread solution for electricity storage are too expensive and not appropriate for large-scale applications. A rough estimation shows that a greater investment (e.g., four to five times) is required for batteries as compared to renewable power plants if we wish to reach 100% renewable supply. Therefore, there is a need for more cost-effective yet reliable large-scale storage solutions. This is where MES technologies come in. These technologies come in a variety of concepts, configurations, and designs. Although many of these systems look promising, almost all of them have disadvantages. This leads to the conclusion that MES systems should be classified in terms of appropriateness for specific applications, meaning that any of these can be a more suitable solution than others in certain conditions. Therefore, there is not really a strong priority among MES systems until very detailed information about the specific case of application of the energy storage system (including the geographical circumstances, availability of resources, type of energy demand, local energy pricing regulations, etc.) is available. Apart from this, the other conclusion we've come to is that none of the existing MES systems can be considered as a perfect solution and thus there is a need for new technologies with new approaches. In this regard, coproduction or tri-production energy storage systems could be extremely interesting, as the global energy systems are moving towards a smart concept in which all the energy sectors are going to be integrated. Lower cost, larger energy density, not being restricted to topologies, greater roundtrip efficiency, and longer life cycle are the further critical parameters to be considered when designing a new MES system.

The final point is that superior MES systems can generally be implemented anywhere for any energy system. Special designs of MES systems for very specific applications (e.g., an innovative solution to accompany an offshore wind turbine) are, of course, always welcome.

References

[1] B. Zalba, J.M. Martín, L.F. Cabeza, H. Mehling, Review on thermal energy storage with phase change: materials, heat transfer analysis and applications. Appl. Therm. Eng. 23 (2003) 251–283, https://doi.org/10.1016/S1359-4311(02)00192-8.

[2] G. Alva, Y. Lin, G. Fang, An overview of thermal energy storage systems. Energy 144 (2018) 341–378, https://doi.org/10.1016/j.energy.2017.12.037.

[3] Y. Tian, C.Y. Zhao, A review of solar collectors and thermal energy storage in solar thermal applications. Appl. Energy 104 (2013) 538–553, https://doi.org/10.1016/j.apenergy.2012.11.051.

[4] S.K. Shah, L. Aye, B. Rismanchi, Seasonal thermal energy storage system for cold climate zones: a review of recent developments. Renew. Sust. Energ. Rev. 97 (2018) 38–49, https://doi.org/10.1016/j.rser.2018.08.025.

[5] A. Arabkoohsar, L. Machado, M. Farzaneh-Gord, R.N.N. Koury, The first and second law analysis of a grid connected photovoltaic plant equipped with a compressed air energy storage unit. Energy 87 (2015) 520–539, https://doi.org/10.1016/j.energy.2015.05.008.

[6] A. Arabkoohsar, L. Machado, R.N.N. Koury, K.A.R. Ismail, Energy consumption minimization in an innovative hybrid power production station by employing PV and evacuated tube collector solar thermal systems. Renew. Energy 93 (2016) 424–441, https://doi.org/10.1016/j.renene.2016.03.003.

[7] M. Sadi, A. Arabkoohsar, Modelling and analysis of a hybrid solar concentrating-waste incineration power plant. J. Clean. Prod. (2018), https://doi.org/10.1016/j.jclepro.2018.12.055.

[8] A. Arabkoohsar, Combined steam based high-temperature heat and power storage with an Organic Rankine Cycle, an efficient mechanical electricity storage technology. J. Clean. Prod. 119098 (2019), https://doi.org/10.1016/j.jclepro.2019.119098.

[9] A. Arabkoohsar, Combination of air-based high-temperature heat and power storage system with an organic Rankine cycle for an improved electricity efficiency. Appl. Therm. Eng. 167 (2020) 114762, https://doi.org/10.1016/j.applthermaleng.2019.114762.

[10] D.E. Elliott, T. Stephens, M.F. Barabas, G. Beckmann, J. Bonnin, A. Bricard, T.D. Brumleve, G.B. Delancey, C.A. Kruissink, G.A. Lane, W.R. Laws, R.F.S. Robertson, J. Schroeder, Working group A—high temperature thermal energy storage. in: E.G. Kovach (Ed.), Thermal Energy Storage, 2013, pp. 11–26, https://doi.org/10.1016/B978-0-08-021724-6.50008-6 Pergamon.

[11] Z. Abdin, K.R. Khalilpour, Chapter 4. Single and polystorage technologies for renewable-based hybrid energy systems. in: K.R. Khalilpour (Ed.), Polygeneration with Polystorage for Chemical and Energy Hubs, Academic Press, 2019, pp. 77–131, https://doi.org/10.1016/B978-0-12-813306-4.00004-5.

[12] M. Mofijur, T.M.I. Mahlia, A.S. Silitonga, H.C. Ong, M. Silakhori, M.H. Hasan, N. Putra, S.M. Ashrafur Rahman, Phase change materials (PCM) for solar energy usages and storage: an overview. Energies 12 (2019) 1–20, https://doi.org/10.3390/en12163167.

[13] B. Kalidasan, A.K. Pandey, S. Shahabuddin, M. Samykano, M. Thirugnanasambandam, R. Saidur, Phase change materials integrated solar thermal energy systems: global trends and current practices in experimental approaches. J. Energy Storage 27 (2020) 101118, https://doi.org/10.1016/j.est.2019.101118.

[14] A. Benato, A. Stoppato, Pumped thermal electricity storage: a technology overview. Therm. Sci. Eng. Prog. 6 (2018) 301–315, https://doi.org/10.1016/j.tsep.2018.01.017.

[15] Stiesdal, Stiesdal Electricity Storage, https://www.stiesdal.com/energy-storage/.

[16] X. Luo, J. Wang, M. Dooner, J. Clarke, Overview of current development in electrical energy storage technologies and the application potential in power system operation. Appl. Energy 137 (2015) 511–536, https://doi.org/10.1016/j.apenergy.2014.09.081.

[17] A. Arabkoohsar, G.B. Andresen, Design and analysis of the novel concept of high temperature heat and power storage. Energy 126 (2017) 21–33, https://doi.org/10.1016/j.energy.2017.03.001.

[18] H. Blanco, A. Faaij, A review at the role of storage in energy systems with a focus on power to gas and long-term storage. Renew. Sust. Energ. Rev. 81 (2018) 1049–1086, https://doi.org/10.1016/j.rser.2017.07.062.

[19] International Hydropower Association, The World's Water Battery: Pumped Hydropower Storage and the Clean Energy Transition, (2018) pp. 1–15.

[20] C. Hauer, B. Wagner, J. Aigner, P. Holzapfel, P. Flödl, M. Liedermann, M. Tritthart, C. Sindelar, U. Pulg, M. Klösch, M. Haimann, B.O. Donnum, M. Stickler, H. Habersack, State of the art, shortcomings and future challenges for a

sustainable sediment management in hydropower: a review. Renew. Sust. Energ. Rev. 98 (2018) 40–55, https://doi.org/10.1016/j.rser.2018.08.031.

[21] J.I. Pérez-Díaz, M. Chazarra, J. García-González, G. Cavazzini, A. Stoppato, Trends and challenges in the operation of pumped-storage hydropower plants. Renew. Sust. Energ. Rev. 44 (2015) 767–784, https://doi.org/10.1016/j.rser.2015.01.029.

[22] A. Slocum, G. Fennell, G. Dundar, B. Hodder, J. Meredith, M. Sager, Ocean renewable Energy Storage (ORES) system: analysis of an undersea energy storage concept. Proc. IEEE 101 (2013) 906–924, https://doi.org/10.1109/JPROC.2013.2242411.

[23] D. Castelvecchi, Flywheels: Spinning Into Control, Freelance Science Writer, https://web.archive.org/web/20140606223717/http://sciencewriter.org/flywheels-spinning-into-control/, 2007.

[24] B. Bolund, H. Bernhoff, M. Leijon, Flywheel energy and power storage systems. Renew. Sust. Energ. Rev. 11 (2007) 235–258, https://doi.org/10.1016/j.rser.2005.01.004.

[25] S.M. Mousavi G, F. Faraji, A. Majazi, K. Al-Haddad, A comprehensive review of flywheel energy storage system technology. Renew. Sust. Energ. Rev. 67 (2017) 477–490, https://doi.org/10.1016/j.rser.2016.09.060.

[26] A.A.K. Arani, H. Karami, G.B. Gharehpetian, M.S.A. Hejazi, Review of flywheel energy storage systems structures and applications in power systems and microgrids. Renew. Sust. Energ. Rev. 69 (2017) 9–18, https://doi.org/10.1016/j.rser.2016.11.166.

[27] A. Arabkoohsar, G.B. Andresen, Design and optimization of a novel system for trigeneration. Energy 168 (2019) 247–260, https://doi.org/10.1016/j.energy.2018.11.086.

[28] M. Budt, D. Wolf, R. Span, J. Yan, A review on compressed air energy storage: basic principles, past milestones and recent developments. Appl. Energy 170 (2016) 250–268, https://doi.org/10.1016/j.apenergy.2016.02.108.

[29] C. Guo, Y. Xu, H. Guo, X. Zhang, X. Lin, L. Wang, Y. Zhang, H. Chen, Comprehensive exergy analysis of the dynamic process of compressed air energy storage system with low-temperature thermal energy storage. Appl. Therm. Eng. 147 (2019) 684–693, https://doi.org/10.1016/j.applthermaleng.2018.10.115.

[30] A. Arabkoohsar, W.K. Hussam, H. Rahbari, Off-design operation analysis of air-based high-temperature heat and power storage, Energy 196 (2019) 117149.

[31] J. Moradi, H. Shahinzadeh, A. Khandan, M. Moazzami, A profitability investigation into the collaborative operation of wind and underwater compressed air energy storage units in the spot market. Energy 141 (2017) 1779–1794, https://doi.org/10.1016/j.energy.2017.11.088.

[32] A. Arabkoohsar, M. Dremark-Larsen, R. Lorentzen, G.B. Andresen, Subcooled compressed air energy storage system for coproduction of heat, cooling and electricity. Appl. Energy 205 (2017) 602–614, https://doi.org/10.1016/j.apenergy.2017.08.006.

[33] 4ward Energy Research GmbH, http://www.4wardenergy.at/de/.

[34] R. Carriveau, M. Ebrahimi, D.S.-K. Ting, A. McGillis, Transient thermodynamic modeling of an underwater compressed air energy storage plant: conventional versus advanced exergy analysis. Sustain. Energy Technol. Assess. 31 (2019) 146–154, https://doi.org/10.1016/j.seta.2018.12.003.

[35] W. Liu, Q. Zhou, D. Dongmei, C. Lu, Q. He, A review of thermal energy storage in compressed air energy storage system, Energy (2019) 115993. https://kopernio.com/viewer?doi:10.1016/j.energy.2019.115993&route=6. (Accessed 20 December 2019).

[36] W. He, J. Wang, Optimal selection of air expansion machine in compressed air energy storage: a review. Renew. Sust. Energ. Rev. 87 (2018) 77–95, https://doi.org/10.1016/j.rser.2018.01.013.

[37] A.S. Alsagri, A. Arabkoohsar, H.R. Rahbari, A.A. Alrobaian, Partial load operation analysis of trigeneration subcooled compressed air energy storage system. J. Clean. Prod. 238 (2019) 117948, https://doi.org/10.1016/j.jclepro.2019.117948.

[38] H. Guo, Y. Xu, Y. Zhang, Q. Liang, H. Tang, X. Zhang, Z. Zuo, H. Chen, Off-design performance and an optimal operation strategy for the multistage compression process in adiabatic compressed air energy storage systems. Appl. Therm. Eng. 149 (2019) 262–274, https://doi.org/10.1016/j.applthermaleng.2018.12.035.

[39] Y. Zhang, Y. Xu, H. Guo, X. Zhang, C. Guo, H. Chen, A hybrid energy storage system with optimized operating strategy for mitigating wind power fluctuations. Renew. Energy 125 (2018) 121–132, https://doi.org/10.1016/j.renene. 2018.02.058.

[40] G. Dib, P. Haberschill, R. Rullière, Q. Perroit, S. Davies, R. Revellin, Thermodynamicsimulation of a micro advanced adiabatic compressed air energy storage for building application. Appl. Energy 260 (2020) 114248, https://doi.org/10.1016/j.apenergy.2019.114248.

[41] A. Scognamiglio, G. Adinolfi, G. Graditi, E. Saretta, Photovoltaics in net zero energy buildings and clusters: enabling the smart city operation. Energy Procedia 61 (2014) 1171–1174, https://doi.org/10.1016/j.egypro.2014.11.1046.

[42] A. Dietrich, Assessment of Pumped Heat Electricity Storage Systems through Exergoeconomic Analyses, E-Publishing-Service der TU Darmstadt, 2017.

[43] A. Benato, Performance and cost evaluation of an innovative pumped thermal electricity storage power system. Energy 138 (2017) 419–436, https://doi.org/10.1016/j.energy.2017.07.066.

[44] Energy Storage Association, https://energystorage.org/why-energy-storage/technologies/pumped-heat-electrical-storage-phes/.

[45] A. Arabkoohsar, An integrated subcooled-CAES and absorption chiller system for cogeneration of cold and power. in: 2018 International Conference on Smart Energy Systems and Technologies, 2018, pp. 1–5, https://doi.org/10.1109/SEST.2018.8495831.

[46] A. Arabkoohsar, G.B. Andresen, Thermodynamics and economic performance comparison of three high-temperature hot rock cavern based energy storage concepts. Energy 132 (2017) 12–21, https://doi.org/10.1016/j.energy.2017.05.071.

[47] A. Arabkoohsar, H. Nami, Thermodynamic and economic analyses of a hybrid waste-driven CHP–ORC plant with exhaust heat recovery. Energy Convers. Manag. 187 (2019) 512–522, https://doi.org/10.1016/j.enconman.2019.03.027.

[48] Siemens, Siemens High Temeprature Heat and Power Storage Project, https://www.siemens.com/press/en/pressrelease/?press=/en/pressrelease/2016/windpower-renewables/pr2016090419wpen.htm&content=WP, 2016.

[49] T. Morstyn, M. Chilcott, M.D. McCulloch, Gravity energy storage with suspended weights for abandoned mine shafts. Appl. Energy 239 (2019) 201–206, https://doi.org/10.1016/j.apenergy.2019.01.226.

[50] A. Berrada, K. Loudiyi, I. Zorkani, System design and economic performance of gravity energy storage. J. Clean. Prod. 156 (2017) 317–326, https://doi.org/10.1016/j.jclepro.2017.04.043.

[51] R. Cazzaniga, M. Cicu, T. Marrana, M. Rosa-Clot, P. Rosa-Clot, G.M. Tina, DOGES: deep ocean gravitational energy storage. J. Energy Storage 14 (2017) 264–270, https://doi.org/10.1016/j.est.2017.06.008.

Index

Note: Page numbers followed by *f* indicate figures, and *t* indicate tables.

Printed in the United States
By Bookmasters